The Pattern of Evolution

The Pattern of Evolution

Niles Eldredge

 W. H. FREEMAN AND COMPANY
NEW YORK

Cover image: Photograph of a fossil ammonite by
                John Cancalosi/Tom Stack & Associates.

Cover design: Victoria Tomaselli
Interior design: Diana Blume

Library of Congress Cataloging-in-Publication Data

Eldredge, Niles.
    The pattern of evolution / Niles Eldredge.
        p.   cm.
    Includes bibliographical references and index.
    ISBN 0-7167-3046-4
    1. Evolution (Biology)   I. Title.
QH366.2.E535   1998
576.8—dc21   98-34754
                                                                              CIP

Printed in the United States of America

Second printing, 1998

To James Hutton, first to show us that patterns in the earth's crust reveal both history and dynamic processes; to George Gaylord Simpson, who showed us that patterns in the history of life have much to teach us about how the evolutionary process works; and to Marjorie Grene, who predicted that hierarchy theory would do good work.

Contents

Acknowledgments

I am deeply grateful to my editors, seriatim, Jonathan Cobb and John Michel for their enthusiasm for this project and for their fruitful ideas. Thanks also to Ms. Portia Rollings for her timely and expert preparation of the illustrations.

CHAPTER 1

Ode to *Cecropia*: Discovering Nature

Somewhere in the dense foliage of the nature preserve off Route 3 in the eastern town of Humacao, Puerto Rico, there is a *Cecropia* tree, instantly recognizable by its spindly gray trunk, bamboolike rings, and large lobed leaves. I had barely noticed the tree as my wife and I passed it a half hour earlier on our way to the Caribbean shoreline. On the way back, though, while we tried to spot the endangered Puerto Rican woodpecker we had heard gently tapping nearby, the matter was entirely different. Confronting that tree triggered a paroxysm of confusion and self-doubt in my mind, as if it had seized up and could no longer think.

"My God, Richard Dawkins must be right after all!"

It was an extremely dark and unwanted first thought. I struggled to understand what was going on in my head. Dawkins, a biologist of Oxford and famed as the author of *The Selfish Gene*, insists that evolution is strictly the outcome of competition among genes—a competition to ensure, and if possible to enhance, their presence in the next and succeeding generations. Nature is organized around the constant struggle for perpetuation among genes (or at least the organisms that house those genes, according to Dawkins), a vision that once led him to predict that one day, even the complex structures of ecosystems would be understood as a relatively direct and simple by-product of those struggles. In contrast, I have had the habit of pointing to that very passage from his writing as the most extreme, illogical, and indefensible of all his evolutionary claims.[1] Yet here I was, on a pleasant, calm birding excursion with my wife, having my brain shout out at me: "Richard is right!"

Why would I suddenly be plagued by such a monumental switch in thinking just by walking by a single tree? *Cecropia* trees of several different species commonly occur throughout the tropics.[2] They are among the most obvious of the plants that colonize open patches in the tropical forests. The phenomenon is known in ecology as *succession*: Pioneer species get into clear areas first, exploiting the great quantities of sunshine hitting the forest floor, and only gradually ceding the space as

1

other species eventually invade, take over, and build a mature, climax forest community. Pioneer species like *Cecropia* have reproductive strategies that allow them to get to these clearings fast: In the case of *Cecropia*, the seeds are actually lying dormant all over the place, just waiting for the opportunity to sprout and take over an entire clearing. So, according to my dark vision, the reproductive adaptations of *Cecropia*—or any other pioneer species, and by simple extension, of all species—must really be about the perpetuation and spread of genes. Worse, the notion of competition for genetic/reproductive success evidently underlies the ecological phenomenon of succession—and, once again by simple extension, the structure of entire ecosystems. Dawkins was right, even in his most extreme claims: The gene's desire to reproduce itself is at the heart of all ecological development! I literally felt ill.

But as we strolled along, I was lucky enough not to catch sight of the woodpecker, and reason began to take hold. True, certain reproductive adaptations that allow pioneer species to find and exploit new opportunities are part of the arsenal of the first species in when succession begins. But so too are a host of *nonreproductive* "economic" adaptations. Foremost among them, insofar as plants are concerned, is the capacity to grow very fast. *Cecropias* can grow as much as 8 feet in a single year, and their large and abundant leaves provide an extremely effective and efficient chlorophyll-laden surface to trap all that sunlight as their photosynthesis swings quickly into high gear.

It has been a persistent theme in my thinking for the past 15 years that organisms do two, and only two, basic things: They obtain the energy they need to develop (for example, as plants and animals do from a fertilized seed or egg), grow, and simply stay alive; and they, typically if not universally, reproduce. So the difference between a guy like Dawkins and a guy like me is really rather simple: Dawkins says in effect that, from an evolutionary perspective, you[3] may eat to live but you live to reproduce; energy procurement is just a necessary condition to allow genes to perpetuate themselves.

I take a more balanced view: You eat to live. You may, as well, reproduce, though your own survival does not depend upon it. I also say that how successful you are at getting the necessary energy to live will, on average, affect how successful you are at reproducing. This is a simple way of restating Darwin's classic formulation of natural selection. Thus, I see natural selection as the fallout of relative economic success on reproductive success: You compete for energy resources in order to live, which may well have implications for the survival of your genes into the next generation. Dawkins would see it the other way: You compete for energy resources because your genes are competing with one another for propagation.

So I had come back to a more balanced view, one far more to my liking: *Cecropia* trees have economic (that is, energy-procuring) anatomical and metabolic adaptations allowing rapid growth as pioneers in addition to the reproductive adaptations that allow them to colonize clearings efficiently. It once again became a matter of perspective—an emphasis on reproductive competition as the prime mover versus my preference for seeing the economic activities of organisms as separate from, and equal in importance to, their reproductive activities. Likewise, it became a matter of seeing that evolution flows from the effects of relative success in one sphere (economics) on the other (reproduction).

But I still wasn't entirely happy. Just as we reached the parking area of the nature reserve, a gorgeous male Puerto Rican woodpecker flew across our path, alighted in a nearby tree, and offered a beautiful show. Nice. But something was bugging me still, something really *obvious* that I was missing.

Then suddenly I saw why my mind really did boggle 10 minutes earlier at the *Cecropia* tree. The day before we had visited the only tropical rainforest of the United States National Forest system—the Caribbean National Forest, better known by the name of the mountain it sits upon: El Yunque. We had stopped to take in the newly opened visitors center. The introductory film covered some of the basic biology and climatic statistics of the rainforest—and reminded us that Hurricane Hugo had greatly damaged El Yunque in 1989. In fact, the rainforest had taken such a hit that El Yunque is generally credited for sapping Hugo's strength—thus sparing the rest of Puerto Rico from even heavier damage.

Midway up El Yunque's slopes we stopped at Yokahu Tower, an observation tower built by the Civilian Conservation Corps in the 1930s. The view of the sea and surrounding countryside was fabulous. But what really caught my eye were the prodigious stands of—you guessed it—*Cecropia*. A lighter green than the mature rainforest, dense swards consisting entirely of *Cecropia* were everywhere, especially as we looked east from the tower. The devastation of Hugo was amazing—a devastation still plain to see a little over 8 years later. A devastation marked by *Cecropia*.

In retrospect, it is amazing that I should have seen that lone tree as a threat to, rather than confirmation of, the picture of evolution I have spent the past 30 years developing. For the mental events of my *Cecropia* parable took place, not way back in the '60s and '70s, during the early days of infighting with the ultra-Darwinians, but rather in February 1998—after I had already finished a preliminary draft of this very narrative. This is a book that asks the question: How has evolutionary theory, involving ideas on how biological evolution happens, remained so disconnected, so aloof, from the domain of the rest of science—the world of matter-in-motion[4] studied by physicists, chemists, and earth scientists?

This narrative is a search for that very connection, one that asks how patterns in the history of the earth and of life can reveal what that connection must look like. And, by the final chapter, I propose an answer: The connection between the evolution of life and the physical history of the earth lies through ecology. I make the very strong claim that nothing much happens in biological evolutionary history until extinction claims what has come before. And species extinction is an ecological phenomenon: the fallout of ecosystem disruption, degradation, and eventual outright destruction.

How could I have fallen so blindly into that momentary switch to the Dawkinsian camp? This camp, with its conviction that everything from evolutionary history to the structure of ecosystems emanates from the competitive urges of genes, if anything isolates evolutionary biology still further from the physical realm of matter-in-motion.

Ecological succession does not happen without physical destruction of mature ecosystems. This is true whether we're talking of fires in grasslands and forests, of storm damage to coral reefs, of sandbars slowly traversing a bay and wiping out the invertebrate fauna living in and on the sediment surface, or of a hurricane hitting a mature tropical rainforest. For *Cecropia* to sink its roots in, a clearing has to be made. And there is no way a clearing will appear through the simple lives and deaths of individual organisms. It takes an event, a physical event, to trigger succession.

Ecological succession—where pioneer species colonize degraded habitat, eventually to be replaced as other species come to be re-established—is a process that produces pattern. Hurricane Hugo's effect on El Yunque was an isolated event. But other such events, varying only in their intensity and coverage, produce extremely similar results. Over and over, through the entire tropics of the western hemisphere, the same pioneer species (such as my much-beloved *Cecropia*) show up first and are in due course followed by an orderly succession that differs only in detail.

Patterns in the natural world are extremely important. As we shall see, they pose both the questions and the answers that scientists formulate as they seek to describe the world: the nature and behaviors of entities such as atoms and continents, or organisms, species and ecosystems, according to one's particular bent and expertise. Science is a search for resonance between mind and natural pattern as we try to answer these questions. You look from the Yokahu Tower at the *Cecropia* stands on El Yunque and you know some devastating force destroyed large sections of the rainforest not all that long ago just as sure as if you were looking at the bits of greenery that began peering through the ash on Mount Saint Helens a few years after its eruption. You might not be able to tell if this force was a tremendous wind causing blowdowns, or a fire (though a practiced eye surely could), but you know something set the ecological clock back to midnight, kicking off the standard reactions of normal succession.

So, too, with large-scale, deep-time patterns in earth and biological history. Patterns in the history of life have been suggesting for at least a half century that there are regularly occurring sets of conditions that seem to control evolutionary activity—dampening it, often for longer periods, and triggering often rapid evolution at other times. Like ecological succession, evolution produces, not isolated events, but repeated patterns which hold clues to how the process works—specifically, how the physical world of matter-in-motion impacts the biological realm.

There are two morals here: First, short-term events, like ecological succession, produce well-known characteristic patterns, patterns that reveal the very nature of the process; second, so too do larger scale evolutionary events. And while evolutionary biologists willingly concede that isolated events affect the course of evolution (as when the dinosaurs became extinct when the earth was struck by one or more comets 65 million years ago), they have been slow to acknowledge that these sorts of events—all physically induced—produce repeated large-scale historical patterns that are just as regular, just as lawlike, as ecological succession. The second moral—that all such ecological and evolutionary events are triggered by external, physical factors—is especially delicious. It suggests a metatheory (developed in the final chapter) that links such short-term ecological processes as succession with various other intermediate and truly long-term global ecological and evolutionary patterns.

Finally, seeing that everything from ecological succession through speciation and the radiation of large clusters of species (for example, "higher taxa," such as the class Mammalia) is triggered by physical, usually destructive events means one other thing: I needn't have worried. Dawkins's "selfish genes" are as incapable of triggering ecological succession as they are of directly causing evolutionary history.

Ten minutes walking in the Humacao woods, replete with more than one Gestalt switch, not to mention a final epiphany, reveals more than just the nature of the evolutionary process, it reveals the very basic way we all think. We encounter pattern; it impacts us, sometimes wholly subliminally; we wrestle with that pattern; and sometimes we end up seeing nature differently than when we began.

My experience in the Humacao rainforest stands as a microcosm of this narrative. While the ultimate goal is to connect biological phenomena with the physical world, I am also taking a hard look at scientific creativity, especially how the mind grapples with patterns in the natural world. The path I follow (studded with *Cecropias* lying in wait) is historical. It begins with the fragmentation of early professional science (in the late eighteenth and early nineteenth centuries)—a necessary step to focus on the particulars of organization and history of both the earth and of life. I ask why was it that the early geologist James Hutton could discern so clearly physical process from historical pattern in the rocks he studied, yet his successors quickly divorced the study of process from the contemplation of historical pattern? Why, despite Hutton's obvious successes, did it take geology 100 years longer than it took biology to develop a coherent evolutionary theory of the earth? And what patterns in the rock record, including the fossilized remnants of life's history, led to the development of the geological timescale, and to an accurate grasp of the enormous depth of geological time?

In the biological realm, what major pattern did Carolus Linnaeus see that led him to his groundbreaking system for classifying all living things? And why did it not lead him (as it later led Charles Darwin) to postulate a biological theory of evolution? On the other hand, what pattern did Jean Baptiste, Comte de Lamarck, who did entertain evolutionary notions, see in the living world, and why did he (largely) fail to convince his contemporaries that life had evolved? And what *were* the patterns that Darwin eventually saw—and which patterns did he not acknowledge—whether through design or not?

Returning to geology, what patterns in the earth's crust motivated Wegener to propose "continental drift?" Why was he discredited, only to have his ideas resurface in fundamentally similar form when "plate tectonics" was formulated by geophysicists in the mid-twentieth century? What is the relation between biological and physical geological pattern in the history of debates over the geological evolution of the earth?

What patterns in the early data of genetics temporarily threw Darwinism into eclipse—and what resolved these apparent conflicts? What evolutionary pattern did Theodosius Dobzhansky and Ernst Mayr see that Darwin ignored? How, at last, did George Gaylord Simpson, 150 years after Hutton, show that historical evolutionary patterns can indeed reveal something about the nature of the evolutionary process? And what does the idea of Eldredge and Stephen Jay Gould, "punctuated equilibria," have to do with the patterns and perceptions of Dobzhansky, Mayr, and Simpson?

Finally, what other evolutionary patterns are there? How do they connect with ecological patterns, both large- and small-scale? And how do patterns in the history of the earth mesh with those in the history of life? What is the connection between the physical realm of matter-in-motion and biological evolution?

What, in the end, drives evolution?

As the answers to these and many other questions unfold, we begin to converge on a coherent theory that links the evolution of life with the physical history of the planet—not as a long series of isolated events, but in regular, repeated, lawlike patterns that can be generalized into a coherent theory of physical and organic evolutionary process. Along the way, we also see how process is inferred from pattern—the fundamental ingredient of genuine scientific discovery.

We will need a few analytic tools to take with us. We need to settle the question: Is there any meaningful distinction between experimental and historical science? We need to be real clear about what the word *pattern* means. And we need to understand the distinction between reductionism, the predominant mode of scientific explanation of the twentieth century, and what is, in many ways (if not exclusively so), its antonym: hierarchy theory. Thus armed, we can begin a systematic search for *Cecropia* trees along the road of scientific history.

History Matters

Science as a collective enterprise has gone far toward an accurate depiction of what philosopher Mario Bunge once called the *furniture* of the universe—the categories of entities that exist.[5] We also know a lot about the interactions between such entities—what the furniture is for. It is this interactive behavior among entities of the material world that produces the actual historical events of the universe. And that is what science is: the description of the material things of the universe and the interactions among them, a characterization often attributed to Ernst Mach.

We learn about the nature of these entities, and their manner of interaction, by studying historical events. This is as true of subatomic physics as it is of paleontology, though the point is sometimes missed, even by some of my more renowned colleagues in biology and geology. Famed Harvard (once American Museum of Natural History) ornithologist and evolutionary biologist Ernst Mayr, for example, drew a distinction between what he called *functional* and *historical* science. So did the great paleontologist George Gaylord Simpson, who likewise left the American Museum in New York to end his career at Harvard.[6] Among the "nonhistorical" sciences, such as physics and chemistry, Simpson liked to speak of the search for *immanence*, the immutable laws of nature. In contrast, for both men, historical sciences such as geology, paleontology, and evolutionary biology in general, focus on the interpretation of individual historical events—events destined never to be repeated as time marches on.

Such distinctions tend to make younger practitioners in historical geology and evolutionary biology cringe. The real distinction, they fear, is between the "hard" (or even worse, the so-called real) sciences—such as the aforementioned physics and chemistry, together with their straightforward applications to biology (biochemistry, molecular biology) and geology (geophysics, geochemistry)—and the "soft"er disciplines of historical science. Indeed, if historical science is solely about recounting the singular facts of history—of the separate lineages of organisms over the 3.5 billion years history of life, or the geological history of the North American continent—what is there about the enterprise that actually makes it "science" at all? Apart from the fact that the subject matter is the history of elements of the material universe—in contrast, say, to the history of languages, nation-states, or nineteenth-century cornets— where are all the accustomed trappings of the scientific enterprise? Where, for example, are the controlled experiments, repeatable over and over with modified boundary conditions, where general statements can emerge such as those denoting the complex steps in the process of manufacturing proteins according to DNA blueprints?

The malaise that Simpson, Mayr, and others caused, or at least mirrored, with their distinction between functional and historical science had some definite effects, especially on the practice of evolutionary biology in the 1970s, 1980s, and 1990s. Some were positive, others not so. On the positive side, Simpson's characterization of taxonomy (the field of biology concerned with naming and classifying species) as an art[7] became the battle cry in a famous and highly successful effort to formulate strict rules for the reconstruction of evolutionary relationships. This so-called phylogenetic systematics (or *cladistics*, as it is commonly known) has by now reached global routine acceptance as the only rigorous, repeatable, and testable—hence truly "scientific"—way of deriving evolutionary histories, and taxonomies, of the world's 10 million or more species.

Ironically, the core procedure of cladistics is identical to the precepts worked out earlier in historical linguistics—and still earlier in the relatively obscure field of stemmatics.[8] Stemmatics is the elucidation of the history of manuscripts, especially ones that had undergone extensive copying histories, such as medieval illuminated manuscripts. In principle, photocopying yields limitless utterly perfect copies; there is no chance for introduced error, no happenstance conversion of a letter here or there. Not so when human scribes have been doing the copying: The handwritten version of typographical errors—generally adding an element of chaos, but sometimes creating new, and originally unintended, meanings—occasionally shows up. As manuscripts have been disseminated to other libraries, there to be copied still more, the errors of individual versions have been passed along. Recognition of lineages marked by the handing down of old errors and the accumulation of later errors exactly mirrors the biological process of *descent with modification*—Darwin's original phrase for what has ever since been known as evolution. As time goes by, as lineages diversify (like manuscripts going their separate ways), and as evolutionary innovations and modifications (like typos) occur, a distinctive, layered history of modification accumulates—one that can readily be read backward. Ancient lineages stand revealed as collateral kin when ancient "errors" common to them both emerge.[9]

But if systematic biology has become more rigorous, the simple fact that its rules of procedure remain virtually indistinguishable from stemmatics and historical linguistics, coupled with the realization that the aim of all three disciplines is an accurate genealogical analysis, only serves to strengthen the Mayr–Simpson dichotomy between functional and historical science. Where is the search for generality in stemmatics, historical linguistics, and phylogenetic systematics? If such there be, generality lies in the realm of ever-widening confluences of lineages—of recognizing that thousands of manuscripts all owe their origin to a single work, or that Indo-European is a unified family of languages, or that Eukaryota consists of all species—protoctistans, plants, fungi, and animals, united by cellular features such as a discrete, chromosome-bearing nucleus bounded by a double-layered wall. And this is a very different sort of generality than $F = MA$ or $E = MC^2$. From this perspective, the scientific status of cladistics devolves strictly from its subject matter, its concern for the genealogical relationships of species, members in good standing of the material universe.

If sheer rigor is insufficient to blur the distinction between functional and historical science, we might consider an alternative gambit meant, however unconsciously, to remove the taint of historicity. I have in mind events of the past thirty-five years or so in evolutionary biology. Here history as a tool for explaining how life evolves has been, in effect, denied, shunted aside in favor of purely genetic and molecular descriptions. Geneticists and paleontologists are still very much at each other's throats. We paleontolgists are still accusing geneticists of ignoring history, particularly patterns that might shed light on how the physical world regulates the expression of natural selection. Geneticists for their part still refer to paleontologists as "phenomenologists," quaint in our persistence at trying to wring some sense of how the evolutionary process works by looking at the long-dead products of evolutionary history. Fossils, conceded by all to have once been alive, to have had genes that they inherited from previous generations and passed on to succeeding ones, alas no longer have anything directly to tell us of such matters. Because we learn of the mechanics of genetic transmission from creatures now alive, the false corollary is often drawn that only geneticists armed with laboratory and field data on the transmission of heritable information are in a position to say anything about the mechanics of the evolutionary process.

Thus evolutionary biologists, primarily if not exclusively those who study the genetics of entire populations, see themselves firmly allied with the "functional" scientists. Indeed, I have accused them of being so smitten with "physics envy" that they have committed what I believe to be an unpardonable sin of portraying natural selection as an active process rather like, say, gravity. Natural selection, to a modern ultra-Darwinian, is the competition among organisms (or even among their genes, a notion we owe primarily to William Hamilton and later to fellow Oxfordian Richard Dawkins) for reproductive success—an active race to leave more copies of one's own genes to the next generation.

What is good about the ultra-Darwinian gambit is that, in their identification with functional science, they feel free to search for generalizations about how evolution works. Much as I feel their enterprise to be substantively wrongheaded, I do applaud their explicit search for general statements about how things work—in keeping with the general sense of what science ought to be all about in the first place. But in factoring history *out* of the equation, ultra-Darwinians have created a demonstrably shallow picture of the evolutionary process.

So we need a third gambit, one that admits history as fair game to the scientific enterprise—but does so without exclusive attention to the individual facts of the matter. One that focuses on historical pattern. To that end, we need to return first to an idea introduced earlier: that, if considered carefully, it is literally the case that all science, from the ephemera of subatomic physics to the depths of paleontological time, is bound up in the analysis of historical events—the outcomes of what the material things of the universe actually do.

What, for example, could be farther removed from paleontology—an "historical" science par excellence—than the physics of subatomic particles? For most of the twentieth century, it has been the branch of science that managed to surpass all others in its cachet, its aura of the very quintessence of science. Yet nuclear physics is a gritty mix of theory and experimentation—and the experimentation consists of the detection of particles released in subatomic collisions. Part of the challenge is for theory to predict what ought to be observed if certain theoretical postulates on the mass and behavior, the nature of subatomic particles are correct. The other part of the challenge is for theory to explain unpredicted results.

But what are these results? Ever since the first crude bubble chambers were designed, experimentalists have improved ways to make the trajectories of subatomic particles visible. It is the evanescent flash, the lingering trace of the telltale path, that enables physicists to say something definitive about these elusive particles. The more such experiments can be repeated, the more such identical trajectories can be documented, the more confident physicists can be of the nature of these minute entities—whether or not they appear to agree with predictions arising from theory.

The streak of a subatomic particle flying through a cloud chamber, frozen on a photographic plate, is a pattern. One such exposure is tantalizing, but it is the repetition of the results that builds confidence in the reality of such a pattern.[10] Once confronted with essentially similar results, physicists can either congratulate themselves on theory confirmed or return to the chalkboard to try to puzzle out the unexpected.

Simpson was surely right when he contended that scientists such as our subatomic physicists are after the "immanent" properties of the universe—in this instance, of the tiniest of building blocks of atoms. But he was wrong, on two counts, when he thereby concluded that such work differs in principle from his own in paleontology: he was wrong to suggest that physicists don't consider historical data; and he was wrong to suggest that paleontologists do not look for patterns—repeated events of haunting similarity—in their own search for immanent properties of the natural world. As a fellow paleontologist, much younger and following in Simpson's footsteps, I see his work as actually epitomizing the search for functional, dynamic meaning in repeated patterns in the history of life.

That physicists do not "care" about the fates of individual particles except as they reveal their general properties simply means that no one has yet seen a use for a strictly historical component of physics. A particular history of a single water molecule over 3 billion years since its advent from outgassing of a long-gone volcano is of no particular intrinsic interest. Yet it does not follow that the data of physics are ahistorical. It is obvious that all phenomena, however brief, have a temporal component and that it is the behavior of entities of the material universe over stretches of time—be they nanoseconds or billions of years—that provides the human mind with an opportunity to grapple with the "furniture" of the universe. Nor is it even true, strictly speaking, that physicists completely eschew history in their take on the universe: For example, take the much-publicized search for a unified field theory, wherein gravity, electromagnetism, and the so-called strong and weak forces are united in one general theoretical statement that "explains everything." That theory is also popularly imagined to be a true description of the nature and the events of the first 10^{-43} seconds of the universe. Everything is imagined to have been, at that moment, literally united, in a world before atoms, nuclei, and even subatomic particles took on specific, corporeal existence.

The history and fates of individual biological elements can clearly be seen to have their own intrinsic interest. The fascinating history of the gene that codes for hemoglobin is checkered with odd, noncoding segments still associated with the gene which record a shift in locus of the coding site. Yet the great interest with this history really lies in what it tells us, through one well-worked out example, about how genes actually work: what their relation is to their coded end-product proteins.

Or consider the history of another, larger-scale biological entity: our own species, *Homo sapiens*. There is perhaps no subject in all of biology more inherently fascinating to more people than the story of who we are, where we came from, and how we fit into the material world around us. Here is focus on a single case history taken perhaps to the highest degree. Yet it remains true that at least part of our inherent fascination in human evolutionary history lies in discovering to what degree we, as a species, conform to the basic properties true of all sexually reproducing species—and the degree to which we differ. In other words, we search for general understanding of who we are partly by looking at generalizations about the inherent properties of species, and comparing our own condition with what we take to be the "norm" in mammalian species in general.[11]

There is one genuine oddity of the insistence of both Simpson and Mayr that evolutionary biology is intrinsically different, in its emphasis on history, from the search for "immanence" characteristic of the physical sciences. That peculiarity is that each of them is best known for his contribution to general evolutionary theory—for generalities on how the evolutionary process characteristically works. I defer detailed discussion of their contributions until Chapter 5. However, at this point I must note that Simpson's idea of *quantum evolution* (a theory pertaining to the rapid evolution of large-scale biological groups such as whales or bats), and Mayr's identity with much of the seminal work on *allopatric speciation* (a theory that sees geographic isolation as an essential prerequisite to the derivation of descendant from ancestral species), are quintessential examples of generalized patterns. Such patterns of historical events are encountered over and over in the data of paleontologists and systematists. In the case of quantum evolution, scientists confront patterns of anatomical change in the context of geological time, and in the case of allopatric speciation other scientists see patterns of anatomical change in extant organisms in the context of geography. Indeed, biologist (cum philosopher and Darwin scholar) Michael Ghiselin is dead right when he takes Mayr to task on precisely this point: according to Ghiselin,[12] allopatric speciation is a generalization that is indistinguishable in basic form from any "law" in any field of science.

Repeated historical events, whether over nanoseconds or millions of years, that share a haunting similarity are the patterns—the phenomena, the real data—of all science. It is patterns that pose the questions. And, perhaps counterintuitively, it is also patterns that in many ways suggest the answers—the explanatory hypotheses, the theories—to those questions. Science is a way of seeing the material world, and pattern perception lies at its core.

Perceiving Patterns

In September 1995 I had the privilege of searching out one of the world's rarest birds—the Taita falcon—known from only a few remote cliff faces in eastern and southern Africa. I was in Victoria Falls, Zimbabwe, and had only an hour to spare between obligations as an ecotourism host. But I was fortunate to be in the company of Woody Cotterill, Director of the Biodiversity Foundation for Africa, and Woody had a friend, Kit Hustler, working in Victoria Falls. It was Kit who had rediscovered these elusive birds at the falls just over a decade earlier, and he had been keeping close tabs on them ever since.

Circumstances were not auspicious. It was 2:00 p.m., the prime heat of the day and the very worst time to look for any wildlife. But we had to try. A short walk up the railroad tracks took us to a trail leading to the 1,000-foot-high cliff face. (This trail bordered, I was told, an otherwise obscurely marked and still-live minefield left over from the war that ultimately saw Northern and Southern Rhodesia go their separate ways as

Zambia and Zimbabwe.) The chasm, with the Zambezi a mere glint below, is perhaps a quarter mile across and bounded by sheer walls of
brown-black basalt. Nothing much was stirring as we set up the spotting
scope. As we chatted, keeping an eye out on the cliff face on the
Zambian side of the river, I glimpsed a bird flying against the basaltic
backdrop. It turned out to be a dove. Back to chatting.

Then I saw another bird moving across the rocky wall. As I focused
my field glasses on it, it began to rise, soon soaring over the top of the
cliff and settling in a spindly, leafless tree. Fearful of losing the bird, I
kept my glasses trained on what was really not much more than an indistinguishable, and now motionless, speck, while trying to explain to my
companions just where it was. But before they could get the scope zeroed
in on it, the bird jumped from its perch and glided back down the cliff
face, disappearing into a niche marked with just a little telltale
whitewash—the guano that marks a favored perching spot. I could no
longer see the bird, but Kit knew all the Taita perching sites, and he
soon had the scope trained on the right spot.

And there it was. In a stroke of dumb luck, against all odds, we had
apparently caught a nesting pair changing guard over the eggs; the bird I
had spotted was taking a break. Kit carefully adjusted the focus and
moved aside to let the bird aficionado from far away have a look at this
very-hard-to-find species. I bent over, eager for the crowning, definitive
look, but I could not see the bird. Kit, naturally, asked, "Do you see it?"
And I, of course, not wanting to seem stupid by admitting that I really
didn't, said something lame like, "Well, yeah, I guess so. Is it looking
away from us?" All I could really make out was an empty ledge. All I
could see was basalt.

Woody took a turn, and also (I was relieved to hear) had trouble
making the bird out. Kit checked again, thinking maybe the bird had
flown; but it hadn't. It was still very much there, big as life. Determined
to see the damn thing, I took another turn squinting into the scope's
eyepiece. Still nothing but rock. But then, all of a sudden, I saw it. The
bird just popped out from the background. It had been there all along, I
had been staring at it all the while, yet I could not see it. But after I finally learned how to see it, it became impossible not to—and impossible
to understand how I could not have seen it earlier.

It is one thing not to notice an obscure dark gray and russet speck a
quarter mile away sitting on a vast, rust-stained rock wall—especially if
you are not a birdwatcher. Even if you are, but do not know that Taita
falcons are in the neighborhood, chances are you would miss that bird.
Even Kit, who knows they frequent the area, without seeing the bird fly
first, probably would not have located that motionless bird sitting on the

ledge. But I knew it was there, I wanted to see it, and I knew very well what falcons look like. For there is an unmistakable general similarity to virtually all falcons, with their blunt heads and, more often than not, pronounced dark "moustachial stripes" against a light-colored cheek. Taitas are like miniature peregrines—a little stockier than other falcons. But they are very falconlike in any case.

So I knew it was there, I knew what I was looking for, I really wanted to see it, and still I couldn't see it. Until I saw it. We have all had experiences like this. Psychologists fond of pattern perception usually start out with a simple skeletonized cube, one that most people see as facing down and to the right, but which also can be seen as facing up and to the left: a "Necker cube." Their subjects almost always feel foolish, inadequate when they can't see the other manifestation of the cube. For nearly everyone, sooner or later that second cubic shape will pop out, unbidden. But it is easier to see if you know it is there. Performance anxiety might hinder that first switch; and the second shape might give way, again unbidden, to the first. But after a few minutes' practice, most viewers can summon up both cubic forms more or less at will. It was just this sort of Necker cube switch that hit me like a ton of bricks when I passed that *Cecropia* tree for the second time. Suddenly I could see it "facing up and to the left."

To see something in the natural world, it not only helps, but is apparently imperative, that we have some mental picture of that "something" already there in our mind's eye. Recognition is a matching up of mental pictures and physical reality—a resonance between mind, the world's furniture, and their events, mediated by the senses.[13]

If we need a mental picture before we see something, it is only fair to ask where those pictures come from in the first place. It is easy enough to point to a field guide of southern African birds as the obvious source of my mental picture of a Taita falcon—and my deeper experiences of matching birds with their pictures as I sought and saw falcons in many other places prior to my visit to Victoria Falls. And as rewarding as I may find it personally, I am merely identifying and recognizing an object, an individual bird of a particular species, which is already quite well known to humanity in general: to ornithologists, to impassioned birdwatchers, and undoubtedly as well to the local San and Bantu peoples for whom field guides and Linnaean taxonomy form no part of their cultural traditions.

Where do new ideas—new "intentions," new ways of seeing—come from? Goethe was on to something when he said that we have to learn how to see things in new ways. The mind does not create the natural world, but it does search for new ways to make sense of it, to bring real events and entities into focus, to see familiar phenomena in novel ways, to "make the world visible."

I have long thought of the scientific process as a matching up of mental pictures against perceptions of the world—in a search for absolutely the best, most accurate, description of that world. Patterns impose themselves on us. We have but to open our eyes to see them. But it is clear that something more is there as well: a search for new ways of seeing that actually change the very perception of the world itself. New phenomena, new patterns will then be revealed, things and events that have always been there but not previously been seen as important—or even been seen at all.

It is this two-way street—the search for more apt pictures or ideas to explain natural phenomena and the search for new ways of seeing that cause phenomena to "pop out" like Taita falcons against a cliff face— that together form the resonance between mind and material nature that is the heart and soul of science. The search for more accurate depictions and explanations of phenomena already perceived is where most of the serious day-to-day work of science lies. But it is in the learning of new ways to see phenomena that true novelty and creativity come in. Both are vital and in many ways themselves inseparable. Both involve wrestling with patterns in nature—the explanation of agreed-upon pattern, and the search for new ways of seeing new patterns.[14]

Punctuated Equilibria and Pattern Perception

Do patterns just "exist" in nature and simply force themselves on our consciousness? Or is it true that we have to learn to see—to become receptive to the pattern before we can even notice it, much less make some sense of it? If we have to learn, what sorts of factors enter into the process?

To answer these questions, I have selected four very different kinds of patterns from my own work that led up to the original formulation of the notion of *punctuated equilibria*.[15] The goal of that study was simply to document patterns of evolutionary change in time and space in some lineage where there was sufficient anatomical complexity, an abundance of well-preserved specimens, a nonnegligible time span (I was dealing with 6 to 8 million years), and a relatively broad geographical area (roughly the eastern half of North America).

The first such pattern is purely anatomical, dealing strictly with the organization of lenses on the compound eyes of species of a certain lineage of trilobite.[16] Once that pattern was seen, others emerged when the distributions of changes in the eyes were plotted with respect to geological time and geographical occurrence. These patterns in turn suggested the existence of still another major pattern involving the distribution of species in space and time with respect to patterns of anatomical stability and change. In order, they were:

Pattern 1: The Eyes. I was studying the arthropod *Phacops rana*, a trilobite that is part of a much larger group, the suborder Phacopina. Arthropods generally have compound eyes, typically with tens to hundreds of small, densely packed lenses covered by a translucent corneal sheath, but phacopid trilobites are unusual in that the lenses tend to be much larger, with each lens covered by its own separate "cornea." Since the eyes are not exceptionally delicate, they are usually very well preserved on fossil specimens.

In approaching the problem of description of the variation in the anatomical features of *Phacops rana*, I simply chose well-trodden paths: I devised a series of linear measurements (perhaps overly complex and heavily redundant); then I subjected them to a lot of statistical analysis,[17] on the novel big mainframe computers just coming onto the campus scene in the late 1960s. At the time, nothing terribly interesting emerged from all that measuring and computer analysis.

Fortunately, I also decided to count the lenses in each eye as I measured the specimens. Here, too, I followed precedent—though I was faced with a distinct choice. Closely packed lenses on the curved visual surface of an arthropod eye resemble the cells of a honeycomb—simultaneously stacked and spiraling—or the similar distribution of seeds of a sunflower. Talk about pattern! The lenses spiral either downward to the right or downward to the left, depending on how you choose to look at them. And there was precedence for seeing the lenses on these eyes arrayed spirally: In an important early (1889) paper on these trilobites, paleontologist John M. Clarke, then very much a junior scientist working under the aegis of the famous James Hall, had already described patterns of variation in the eyes of different samples of *Phacops rana*. Clarke, though, had chosen to report the lens counts in spiral arrays forming diagonal rows when projected on a flat surface. Clarke reported that the majority of specimens showed nine such diagonal rows, but there may be as few as 8 or as many as 11. Though as the very smallest specimens had 8 diagonal rows, in general there seemed no rhyme or reason, no overarching general pattern, to Clarke's trilobite eye data.

So I followed, instead, the lead of Euan N. K. Clarkson, who had just finished his dissertation on some similar trilobites in England. Clarkson had seen the other way of looking at lens arrangement in

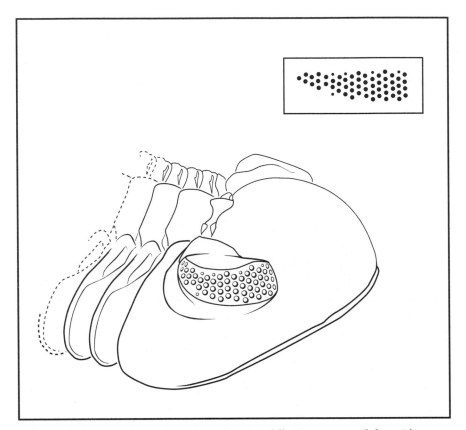

A view of a head and partial thorax of the Middle Devonian trilobite *Phacops rana*. The curved surface of the right eye depicts the lens conformation of a particular specimen reported by John M. Clarke in 1889, as projected on a flat surface within the inset box. The two perspectives reveal both the diagonal rows that Clarke chose to emphasize, and the vertical (dorsoventral) files as seen by Euan N.K. Clarkson. Counting from the front, the dorsoventral file pattern is: 454 545 444 434 232 21 (17, 60).

phacopid eyes: as straight up-and-down vertical columns of lenses, which he dubbed "dorsoventral files." Clarkson would have described the *Phacops* in the accompanying figure in this fashion: 454 545 444 434 232 21 (17, 60), meaning that there were 17 dorsoventral files, that the first (in front, the most anterior) file had 4 lenses, the last 1, and there were a total of 60 lenses in that particular eye of that particular trilobite. I'm not sure just why I chose to follow Clarkson over Clarke, except perhaps that Clarkson was the more modern of the two, and his work incorporated at least some sense of pattern, though I had no idea whether it would help in my own research. And of course, his work had been conducted, successfully, in the pursuit of his doctorate degree—which was my main motivation as well!

It turns out that counting lenses John Clarke's way shows a spectrum of variation in the number of spiral rows of lenses. Counting them Euan Clarkson's way shows something very different: small, immature specimens have some lower number of dorsoventral files (say, 13 or 14)—a number which increases as the animal ages and gets bigger. But sooner, rather than later, the number of dorsoventral files stabilizes at a characteristic number (say, 18), with little or no variation to be found. The trilobites keep growing after that point; they, like crabs, shed their outer skeletons to grow, leaving a succession of molts behind for paleontologists to reconstruct the entire growth sequence. But after a stable upper number of dorsoventral files is reached, that's it—no more dorsoventral files appear, big as the head itself might eventually get.

That was pattern 1: growth and stabilization of dorsoventral files during the lifetime of an individual—and very little variation in the adult complement of dorsoventral files within any given population sample of *Phacops rana*. I remember this pattern fairly leaping off the data sheet pages at me one day. Suddenly I realized that all but the babies had the same number of dorsoventral files for any given sample, but not all samples had the same stabilized number of adult files. The difference among the babies was easily explained by growth, but not so with the adults. And for this I needed no computer: The generalization, simple as it was, just popped into my mind as I stared blearily one warm spring day at the umpteenth data sheet crowded with penciled-in measurements, counts, and notations.

Pattern 2: Evolutionary Change over Space and through Time. My original goal was simply to chart the distribution of evolutionary change in a densely sampled lineage over space and time. So the very next thing I did after my mini-Eureka with the dorsoventral file data was to plot the data, simple numbers, on a set of maps. All of us graduate students were armed with a load of blank maps of North America, left over from a particularly rigorous two-semester course on stratigraphy. I made what amounted to a riffle-book, consisting of five maps conforming to five slices of time, from the earliest days through the end of my 6- to 8-million-year-long wedge of Devonian time.

Bingo! Another pattern leapt out: Populations in the Appalachian basin seemed to be invariably 17 dorsoventral-filed, for almost the entire length of Middle Devonian time, with two major exceptions. In the Midwest, the story was altogether different: For 2 million years or so, the number remained stable at 18; thereafter, for at least another 2 million years, the number was also stable, but it was 17, not 18; for the very last part of my time frame, the trilobites had 15 dorsoventral files as their characteristic population number. I had a choice of at least three different "models" to explain this apparent pattern. My original starting point, common to nearly all paleontologists of the 1960s and before, was to expect—predict, actually—that evolutionary change, whatever it might be, should be mainly slow, steady, gradual, and progressive. Whatever my pattern was showing, it didn't appear to be this mode of expected gradual change.

Another choice, another model available to explain my pattern, was *saltation*, meaning a literal overnight transformation from one state to another. This mode of evolution was much denounced by my mentors, as it was associated with the much-derided German paleontologist Otto Schindewolf and the equally castigated German immigrant geneticist Richard Goldschmidt. At first glimpse, such an explanation fit at least the patterns in the Midwest—where the stable number of 18 suddenly gave way to 17, itself stable until it, too, disappeared, replaced by the number 15.

I was never in danger of going the saltationist route, though Steve Gould, my friend and co-developer of the notion of punctuated equilibria, and I have been accused of precisely that on many occasions. I am, I almost hate to admit, basically rather conservative and driven, at least in part, by a desire to be taken seriously. That has always meant staying within the fold of orthodoxy—all the while, of course, still looking for a better fit between the material world and our descriptions of it. And, in any case, there was a third, far better looking model that seemed to fit the data like a glove—a model that had the virtue of coming from Darwinian orthodoxy. That model is *allopatric speciation*, the splitting up of an ancestral species into two or more descendant species, initially through geographic isolation.

Two additional pieces of information clinched this choice. The two exceptions to the otherwise utter stability of 17 dorsoventral files in the Appalachians both came at the beginnings of the known time range of the 17- and the 15-file forms; the pattern is especially clearest for the advent of the earlier, 17-file form. One sample from the earliest *Phacops rana*–bearing rock strata in New York seemed to have both 18- and 17-file forms in it; a much later sample had 17-, 16-, and 15-file specimens—populations I took to be as intermediates between ancestral and descendant conditions. The 17-file form appeared in New York at least 1½ million years before it showed up in the Midwest.

The other piece of information—at once dispelling the possibility of saltationism and clinching the case for standard allopatric speciation—

was that large chunks of time were missing in the midwestern rock sequences precisely when the changeover from 18 to 17 and from 17 to 15 files occurred. The seas in which *Phacops rana* and the two hundred or so other species lived simply dried up during those intervals. Far from recording episodes of evolution, the change from 18 to 17, and later from 17 to 15, dorsoventral files in the Midwest simply marked a repopulating of the seas. The pattern suggests (though surely does not prove) that the 18-file form became extinct when the midwestern seas first withdrew; when the seas came back, the 17-file from was available and took over the newly reconstituted marine habitat of the continental interior.

In other words, we can map evolutionary change in time and space; we might find our expected, even cherished preconceptions not to be fulfilled. But we nonetheless often, even usually, have some perfectly respectable alternative picture, idea, theory, or hypothesis available that lets us "see" the pattern in the first place—one which enables us to explain it fairly readily in familiar, even conventional, terms.

Pattern 3: Stasis. But then there is hindsight. It is the job of all Ph.D. candidates in the sciences to show they can formulate and carry out original scientific research. And, say what one will, there is the unspoken assumption that positive results are to be expected. I was vastly relieved then to find some form of evolutionary change, some coherent pattern that could lend itself to evolutionary interpretation, when at last the pattern of change in dorsoventral file numbers finally struck me.

But *Phacops* has a lot of features—ones I measured, analyzed, and pondered seemingly endlessly, all for naught. Even the dorsoventral files seemed to remain stable, except for two relatively quick bouts of change. It was no thunderclap, just a plodding, slowly dawning realization that *the absence of change itself was a very interesting pattern.* It was a realization that could come only after the relief tendered by finally finding something positive—some actual evolutionary change to write about.

Stasis is the name Steve Gould and I gave to this pattern of persistent variation with little or no net change over the bulk of a species' lifetime—meaning, in many instances, millions of years. It was by far the most important pattern to emerge from all my staring at *Phacops* specimens. Many other paleontologists had seen stasis—for the most part, not seeing it for what it was. Traditionally seen as an artifact of a poor record, as the inability of paleontologists to find what evolutionary biologists going back to Darwin had told them must be there, stasis was, as Steve Gould put it, "paleontology's trade secret"—an embarrassing one at that.

But Gould has also said that "stasis is data," and indeed it is. The trick to seeing stasis itself as a pattern, as a result, and not a nonresult, required only a shift from equating "evolution" with "change"—of seeing evolution as patterns of histories of separate lineages, each marked by periods of stability and change. Punctuated equilibria simply says that the bulk of most species' histories are marked by stability (that is, little or no accumulation of anatomical change), and that most anatomical change in evolution, assumed to be under the control of natural selection, occurs mostly in conjunction with the actual process of speciation, which for the most part occurs through processes of geographic variation and isolation.

Stasis was always there; it had been spotted and discussed by Darwin's paleontological contemporaries before temporarily dropping out of our tool kit of mental images of evolution. But it took a good, long, hard search for examples of evolutionary change—and a realization that the fossil record, even when well suited to the task, provides few and far-between examples of conventionally expected patterns of evolutionary change—before stasis could pop out, like a Taita falcon, for all to see.

Pattern 4: Species Sorting.   Where do new pictures, new imaginings, of the material world come from? I'll leave this rapid-fire survey of early work on punctuated equilibria and the search for pattern for the recognition of an entirely new sort of pattern—a pattern that so far few evolutionary biologists see, save some forward thinking (and rightminded!) paleontologists.

It is a pattern that literally had to be invented first and only then taken to the fossil record to see if it matched or made any sense. This pattern, *species sorting*, springs from a paradox posed by the initial formulation of punctuated equilibria. The paradox is this: One pattern we perceive in evolutionary history is evolutionary trends—for example, the progressive increase in both the absolute and relative brain size of protohumans over the last 5 million years. The trend begins in the ape range (about 400 milliliters), and culminates with our own brains at about 1400 cubic centimeters (actually, Neanderthals had bigger brains). Under the old explanation of evolution, namely, that natural selection slowly but inexorably modifies features through time in a progressive manner, there is no problem: long-term trends are simply the cumulative effect of truly long-term natural selection. But, with the recognition that most species show no such change throughout their long spans of existence, we no longer have available to us this standard explanation of evolutionary trends. What to do?

Conjecture—or conjure—a new pattern, that's what. So Gould and I conjectured that species are real entities, "individuals" with births, histories, and deaths. In keeping with the premise (and my own predilections) that not only is there never anything completely new under the sun but also that it is good to have respectable intellectual company, I hasten to add that others before us, and around us, for different reasons, were making similar suggestions. If species are construed as real historical entities—with the aforementioned births, histories, and deaths—that in itself suggests the very real possibility that there are factors that bias those births and deaths. And that means that there very probably is a higher-level analogue to natural selection—species selection, or, as I have come to prefer, *species sorting*. The apparent directionality of trends, we suggested, might be the result of, say, differential species survival: Brains may become enlarged (via natural selection, presumably associated with the speciation process), but it is the differential survival of bigger-brained species and the extinction of those with smaller brains that produces long-term directionality of increase in brain size in human evolution.

Or so we posited. The purpose of this section is not to argue the success of this or that postulate but rather to try to pinpoint where ideas—pictures of the material world—actually come from. They come from others, often capriciously; they come out of a choice of pre-existing possibilities; they appear as patterns themselves pop out, patterns that had been in a very real sense invisible (in the case of stasis, because only the inverse was regarded as important); and they come of necessity, often because anomalies, inconsistencies, and paradoxes arise that demand a different picture.

These four examples hardly exhaust the interplay between human-devised models and the patterns of events and entities in the material world. We have to learn to see, but there seem to be many tricks and lateral approaches to the task. Sometimes nature seems simply to impose patterns on our consciousness; other times, we deliberately, almost analytically, invent them. For the most part the interplay is probably more subtle, with even the subconscious working on discrepancies and anomalies, probing for a new way to see what is, in reality, only hinted at and not truly perceived as an actual pattern. Wherever they come from in the mind-nature interplay, patterns are the very stuff of science.

Scientific Description and Analysis: The Competing Roles of Hierarchy and Reduction

Over the years, as a direct outgrowth of my own search for pattern and evolutionary meaning in the history of life, I have become strongly attracted to something called *hierarchy theory*, both as a means of approaching the complexities of the natural world and from the conviction that that world is indeed structured in a hierarchical fashion, meaning, for a first and simple approximation, that the material universe consists of parts and wholes. My attraction to hierarchies came from the thinking alluded to above: that to address the apparent long-term directional component of evolutionary history, the evolutionary trends, punctuated equilibria invited the interpretation of species, not just as collections of similar organisms, but as real entities unto themselves: spatiotemporally bounded historical entities. Real things.

Thus organisms are parts of breeding populations (where natural selection takes place). And populations are parts of species. And species are parts of larger-scale systems—higher taxa, such as genera, families, and so on. Of course, part of this was already familiar to me through my early training and experience, specifically, through the Linnaean hierarchy, a system of ranked categories from species on up through kingdoms, where particular taxa (like the mammals, conventionally designated a class) are a part of some larger system (such as vertebrates, conventionally considered a subphylum). As we shall see in the course of this narrative, a predilection for hierarchical thinking has played a subtle, but ultimately significant, role in the history of evolutionary biology. My purpose in this introductory chapter, however, is not to argue for the "correctness" of one particular view, but to explore the ways in which a particular approach—in this instance, hierarchy—can be brought to bear on the general problem of making nature visible. Hierarchy permeates biology. And it is a major, if understated, component of geology and other scientific disciplines as well.

Philosopher Marjorie Grene (with whom I wrote a book about the significance of hierarchies in understanding the biology of social systems) once commented to the effect that everyone talks about hierarchies yet few do anything about them. She meant that, while it is blindingly obvious to all that, for example, chemical compounds consist of (bonded) atoms, which themselves consist of a nucleus surrounded by an electron swarm, with the particles of the nucleus themselves composed of still smaller bits of matter—such realization has little to do with the day-to-day workings of science. Nor does it regularly lead to additional insight about the nature and organization of the material realm.[18]

But Grene did see that a hierarchical approach to natural complexity was, at least in principle, capable of yielding new insights, of actually "doing work" in the description and analysis of biological systems that play a role in the evolutionary process. For one thing, a focus on systems at a particular level, in the context of seeing them as parts of still larger wholes, as well as wholes composed of still smaller parts, in principle would enhance the accuracy of description of the very nature of those systems. For another, the relationships between entities of higher and lower levels is bound to become clarified as well: how, for example, behavior of lower-level entities affects larger levels systems, and how, in turn, larger-scale systems exert an influence on their lower-level constituents.

I'll give some concrete examples of such upward and downward causation in a moment. First, though, anyone who would press the utility of hierarchy in looking at the natural world must get past the charge of *reification*—literally of "making things"—of characterizing some apparent phenomenon as a "real entity" when, in reality, it is not.

Take, for example, the question, Are species real?, which was briefly encountered above. Darwinian tradition had it that, in fact, species are mere collections of similar organisms, capable of interbreeding, perhaps, but assured of evolving themselves out of existence given the mere passage of time. It is by now notorious that Darwin, in On the Origin of Species, basically destroyed the notion of species as real entities—because species were held in pre-evolutionary times to be separately created and immutable. Darwin literally had to get rid of the idea that species are real entities just to establish the plausiblity of the very idea of evolution. Hence the irony of Darwin's very title—given that he never actually did discuss the "origin of species."

It wasn't until the mid-1930s that the notion of species as real entities reinvaded evolutionary discourse. (That movement will be considered in some detail in Chapter 5.) One of the leaders of this movement, Ernst Mayr, said (in 1942) that of course species are real—why else have a theory of their origins? Yet today we still see a spectrum of opinion on the subject. Many evolutionary biologists either take no stance on the issue or deny that species exist in any meaningful sense. And of those groups of evolutionists for whom species are both real and important, there is genuine disagreement as to what species actually are!

It turns out that at least two major factors determine whether or not a phenomenon is, usefully or otherwise, to be considered "real." One is the utility of the construct itself. In this case, the average population geneticist, concerned with natural selection and genetic drift within populations, does not study entire species and so does not really "need" the concept. To him or her, the nature, even the very existence, of some larger-scale, higher-level entity such as a "species" is generally an irrelevance—and quite legitimately so.

Other population geneticists, however, may be interested in comparing the genetics of different populations within a "species"—tacitly, at least, accepting that a species is some kind of thing. For the most part, such biologists adopt the eminently reasonable stance that species are large-scale, complexly organized "packages" of genetic information, within which there is mating and the exchange of genetic information, beyond which there is little or no such genetic interchange (meaning that species tend not to hybridize with one another as a general rule, however fraught with exceptions such a generalization might be). The latter is a modernized version of Ernst Mayr's famous definition of the so-called biological species concept.

Still other evolutionary biologists look at species as real but focus not so much on the ability to interbreed as on various anatomical or molecular signals that reveal patterns of such interrelationship. This study, known as cladistics, was encountered earlier in this chapter, and is undertaken by systematists interested in the evolutionary interrelationships among a chain of closely related species. Often what seems to a geneticist to be one natural entity, united by an ability to interbreed among all its constituent populations, might seem to a systematist to be two different species. The systematist sees that one part of the geneticist's "species" is actually more closely related to yet another group of populations. Thus occurs what I call the "eternal species wrangle," where even biologists who think that species are in some sense "real" can't get together to decide what manner of beast a species is.

The usual conclusion to such persistent conceptual quagmires is either to deny that species are real or to simply drop the whole thing. In the case of species, there is clearly an alternative: that nature is organized in more than one way at or near the species level; which depiction of the nature of species you prefer depends upon your initial theoretical presuppositions. In other words, both descriptions of what species are might simultaneously be correct. The problem lies not so much with competing views of nature but with the complexity of nature itself. And that is very interesting, because both descriptions are, in proper context, "correct." All that needs to be done is specify just what species concept is under discussion to avoid what otherwise is invariably a welter of confusion.

Thus point 1 about hierarchies and higher-level entities: Scientists working at lower levels seldom encounter higher-level entities; for the most part therefore, they may safely ignore them. And there is a related, second point: It is somehow far easier for the human mind to grasp the existence of and to see as "real" still-smaller structures, things that are parts of wholes, than it is to see as "real" systems that we ourselves are parts of — to see, in other words, the forest for the trees.

Take, for example, the hugest structures known in the material universe. By the beginning of the twentieth century, it was realized that the sun is a star, and that that star is a part of the Milky Way system. But the Milky Way was thought to be the entire universe. Not until the various spirals and pinwheels dotting the telescopic images of the nighttime sky were seen not as odd individual stars, but as immense galaxies, composed of millions of stars, was the Milky Way seen for what (we now think) it is: just one of millions of galaxies. In this domain, great distances render the ultralarge so very small that it took giant telescopes to reveal them for what they really are: the class of biggest entities in the universe.

And galaxies are entities: spirals of dust and stars bound together by common origins and a gravitational field. They are coherent entities, despite the vast, and very nearly completely empty, spaces between each of their component stars. Solar systems, too, are entities for the very same reason. So is our planet Earth. So, arguably, is each of the coherent tectonic plates of the earth's outer crust. So is each rock, and each of its constitutent mineral crystals, and each of that crystal's molecular compound constituents. So is each atom in each molecule. And so forth. As Nobelist Herbert Simon, founding guru of the modern-day hierarchy movement in science, put it, it wasn't until well into the twentieth century that atoms came to be seen, not as the fundamental building blocks of matter, but rather as complex systems consisting mostly of empty space; each is composed of subparts, which in turn are composed of subparts.[19]

As technology improves, it has become easier to probe the domains of the very small. We somehow take it for granted that familiar objects are composed of parts. The human body, for example, is composed of organ systems, organs, tissues, cells, microstructures, and so on. In a very real sense, we already knew that. Even if it took the refocusing of a picture that went back to the ancient Greeks in order to realize that there are constituent parts to atoms, it was a realization that was readily absorbed nevertheless—even though the sociocultural ramifications of nuclear physics remain far from digested.

But where is the analogue of the telescope, or reverse microscope, to let us see larger systems as real, especially earthly systems of which we are parts? Such systems, like species or ecosystems in biology, have boundaries that are messy, problematic, and the source of boundless contention. It took 100 years after Darwin's epochal 1859 book for geology to make enough sense of the anatomy and "physiology" of the earth to formulate an equivalent evolutionary theory of the earth. But after that relatively recent revolution in understanding it became far easier to specify and take as "real" large chunks of the earth, seeing the plates as interacting parts of a churning, dynamic, yet coherent, whole. Plate boundaries are messy, too, but on the whole "tolerably distinguishable."[20]

It just may be that the emotional distance we human observers have toward nonliving physical systems makes it easier to perceive them.

I have so far omitted consideration of those who, not directly confronted with large-scale systems, deny their existence instead of simply ignoring them. No one denies that atoms or galaxies exist, but there are plenty of people who insist that ecosystems—or species—do not exist. They deny the existence of such systems, not because they cannot see them, or simply do not need them in their depiction of nature, but rather because such entities actually stand in the way of their conceptualization of how the material world is structured and how it runs. This is not in itself a bad thing. For if it's true that our choice of phenomena as "real" depends on the context and the use we make of such constructs, it must also be true that denying the existence of such systems may likewise prove useful.

Most instances of denial of the "existence" of an entity or phenomenon come from scientists who are asked to comment on larger-scale entities. It is mostly when looking up at purported higher levels of organization that scientists are likely to rebel. The feeling seems to be that the larger-scale phenomena, or purported entities, might look as if they exist but are in reality nothing special, nothing that cannot be explained strictly as epiphenomena of lower-level processes. In other words, to many scientists, all causation goes upward. The world is the way it is solely because of various forces acting among protons, neutrons, and electrons. This is one sense of the term *reductionism*—in many ways the exact antithesis of a hierarchical viewpoint. Reductionism means many things. To philosopher Ernest Nagel, for example, reduction means the formal translation of the terms of one field of science (say, chemistry) to those of another, more general field (say, physics).[21] But in another, perhaps more colloquial sense, reductionism literally means the description and analysis of entities and processes of some purported higher-level phenomena strictly in terms of lower-level entities and processes.

Such, for example, is the approach taken by the dominant school of evolutionary process theorists in modern biology—the group I have labeled "ultra-Darwinians," and with whom I have been in active disagreement for over a quarter century on a variety of issues. The ultra-Darwinians either take no special heed of species (or other larger-scaled systems, such as ecosystems) or more interestingly, they deny that such exist in any meaningful sense. As we saw under the boughs of the *Cecropia* tree, their gambit is to explain all biological phenomena of whatever scale as a fairly straightforward outcome of competition for reproductive success, whether among individual organisms, or, preferably, among their genes.

Ultra-Darwinians want to defend their level against reduction to lower levels, but they do not want to concede the existence of analogous higher levels. Yet if we take the existence of higher levels seriously, what difference does their existence make to our description of the natural world? Here is one simple example: My colleagues and I have been claiming that the great anatomical stability characteristic of the vast majority of species in the fossil record stems in large measure from the internal organization of species themselves. We also claim that species and populations are by no means the synonyms that population geneticists tend to claim them to be. The mere existence of species controls, or sets boundary conditions on, what can and cannot happen in evolutionary history.[22]

Thus, natural systems generally seem to affect one another, from the bottom up, as any good reductionist would say. (For example: "A species is a group of organisms created by ongoing reproduction of its component organisms.") Such things are relatively easy to see. That higher levels exert equivalent constraints on lower levels is not as intuitively obvious. But consider natural selection. Mutation may continually produce a "bad" genetic variant, say poor eyesight in some Taita falcon chicks. But natural selection will cull those mutants and toss them out, forever damping their frequency.

Thus, at its simplest, hierarchy is really just the consideration of the possibility that larger-scale systems exist and exert an influence on the way things are—just as do easier-to-see, lower-scale entities. It does make a difference how those elements of the natural world are packaged: whether or not they come as parts-of-wholes, and just what manner of relations the parts bear to the wholes, and vice versa.

Simple reductionism, as the traditional siren song of twentieth-century science, just won't work any more. Take a favorite phenomenon of chaos theorists: the waterfall whose stolidly constant basic profile contrasts so mightily with the chaotic jerkiness of its inner flow. Chaos theorists raise the point that order can arise even from such apparently internally chaotic systems: Be that as it may, the overall form of the waterfall is determined by the rate of flow of a volume of water over the lip of the falls; that is determined, not least, by rainfall patterns often weeks before the measured flow and hundreds of miles away from the falls.

The world is a big place. It consists of big and small parts, and they are all connected in various ways. It is the search for this connectivity that ultimately this book is all about. To explore the connections between the organic and the nonliving components of this world, we must be at least willing to consider a great spectrum of scales and kinds of phenomena. We must not simply insist that we know everything worth knowing already, or adopt the attitude that the systems that stand so readily revealed to us are all that we need consider in the search.

The universe consists of several crucial series of parts and wholes: hierarchies. We will be concerned pre-eminently with three as we trace the history of scientific thought on the evolution of and connections among earth and life: an evolutionary (genealogical or genetic) hierarchy; an ecological (economic) hierarchy; and a hierarchy of physical structures of the earth's crust. Elements of others will, at least implicitly, appear: Local parts of regional, and ultimately global, climatic and hydrographic structures will come into play as we near the end of the trail.

Patterns, too, are distinctly evolutionary, ecological, and physical. They, like the elements of hierarchies, also come in small and large sizes. Patterns are traces of the activities of the elements of hierarchically arrayed systems large and small. The journey along the *Cecropia* trail leads us ever closer to a complete description of the elements, these parts of wholes, of the natural hierarchical systems. It shows us the patterns, the traces of activity consistently played out by these pieces of earthly furniture: organisms, ecosystems, species, continental plates. These elements, juxtaposed with the traces of their behaviors, the pattern of history, are the ingredients of a much richer view of the earth and its life. The journey brings us closer to seeing how biological evolution truly is connected with the physical realm of matter-in-motion. Let's hear it for *Cecropia* trees!

CHAPTER 2

Teasing the Strands Apart

Rational contemplation of the material universe is a relatively recent human achievement. For all we can point to the ancient Greeks, for all the remarkably clear-headed *aperçus* of such prescient independent thinkers as Leonardo da Vinci, it really wasn't until the eighteenth century that we see the first real stirrings of concerted scientific observation, experiment, and thought. It was then that sporadic spurts of individual creativity began to coalesce into a coherent stream of systematic investigation.

If critical thinking, the rejection of less accurate pictures of material phenomena for better descriptions, is a linchpin of growth and progress in scientific understanding, then surely the development of a community of like-spirited individuals is crucial to the endeavor. It is simply not reasonable to suppose that any single scientist can ever be fully relied upon to hold with total dispassion preferred descriptions, analyses, and explanations of the natural world. The disputations for which the scientific community is so justly renowned are truly a group-level phenomenon. To this day the major role of a critical mass of investigators remains less an issue of finding sheer "man-hours" available to pursue some massive or particularly thorny problem[1] than of providing alternative approaches to the apprehension of natural pattern: alternative descriptions and explanations of segments of the material universe.

Such communities began to form locally in the late eighteenth and early nineteenth centuries. Science was at first generally the (pre)occupation of educated men with enough time and financial means at their disposal to explore whatever piqued their curiosities. Clergymen, physicians, gentlemen farmers, lawyers, and the independently wealthy were among the early emerging class of scientists. During these early days, science, as paleontologist and historian Martin Rudwick has put it, was pursued by "gentlemanly specialists," before science had developed as a profession in its own right. Adam Sedgwick, arguably Charles Darwin's most formative and virtually only formal mentor in science, was a clergyman and Professor of Geology at Cambridge. Sedgwick's early friend and collaborator, then later rival in geological investigation, Roderick Impy

Murchison (also known for his daring explorations of Africa) was an ex-army man of independent means. James Hutton, their forerunner in the late eighteenth century, and a critical figure in the development of a rational approach to geological phenomena, was a nonpracticing physician gentleman farmer. Hutton was therefore at liberty to spend as much time as he liked on his hobby—the systematic pursuit of Scottish geological phenomena. Through this undertaking, Hutton was able to elucidate critical generalizations about earth history; more importantly, he could also elaborate how that history was to be studied rigorously.

And thus eighteenth-century science was conducted. Educated European males, driven by their curiosity and liberated by their early training and often independent means, began to form not just a loosely structured community, but often actual social organizations: "societies". These groups provided the context that both fostered the rapidly expanding early stages of concerted scientific effort in a benign, cooperative way and sowed the seeds for competition, rivalry, even hatred. Such a competitive atmosphere can engender even more rapid progress in the growth of knowledge than the more congenial, but often more complacent, spirit of cooperation and harmony usually can.[2]

If, by mid-nineteenth century, a class of (pre)professionals had arisen, openly dedicated to the mental exploration of the material world via sensory perception, older notions of what the universe is actually like were by no means immediately dispelled.[3] In particular, there was no distinction between "functional" and "historical" science in the early days. Why? Mostly because the early gentlemanly scientists were thoroughly imbued with the traditional, Genesis-based account of the history of the cosmos, including the supposed brief history of earth and its living inhabitants. The give and take between an emerging set of thoughts on how science should be done and competing accounts of the actual nature of things had to thoroughly resonate before the cosmos could be granted a non-trivial history in the first place. In other words, the very fact of history had to be established before any dichotomy between functional and historical science could emerge. And the realization that earth and life, severally and together, have had prodigiously long histories is perhaps the crowning achievement of late eighteenth- and especially early nineteenth-century science. It culminated of course, in the publication in 1859 of Charles Darwin's *On the Origin of Species by Means of Natural Selection*. And that double discovery was both contingent on, and partially responsible for, development of a key element in the interpretation of patterns in the earth's crust and in the very fabric of living systems: the idea of uniformitarianism.

No Vestige of a Beginning, No Prospect of an End: Hutton, Lyell, and the Emergence of Modern Science

James Hutton (1726–1797) was one of geology's earliest, and certainly most influential, pattern seekers. Hutton saw that patterns in the rocks have much to say about the earth's dynamic processes. More than any other single individual, it was Hutton who developed both a rationale and a protocol for concerted investigation of the earth. Hutton was actually able to predict, then duly find, specific patterns in the earth's crust. These findings enabled him to meld concepts of process—such as sediments accumulating on a lake bottom, or lava flows emanating from fissures, later hardening into what we now call basalts and rhyolites—with study of the historical sequence of events. Yet, ironically, at the same time, Hutton's work sowed the seeds for the eventual distinction, even divorce, between pattern and process. Thus a separation developed between the study of earth history per se and the investigation of the forces that shape the earth and produce that history.

There has been an ages-old distinction between "physical" and "historical" geology. (Until this day, study of these two "separate" subjects constitutes the first and second semesters, respectively, of introductory courses in geology.) Indeed, the origin of such a distinction lies in the realization that the earth has had such a history, that repeated geological patterns recording separate, yet hauntingly similar, events in earth history imply common underlying causes. These causes can be understood in large measure through observation of the world in which we now live. We owe these perceptions primarily and originally to Hutton. Once these observations were articulated and accepted, geologists were free to pursue separate lines of enquiry. Some focused on active physical processes observable in the present moment, others on deciphering the myriad details of our planet's history.

Hutton moved to Edinburgh in 1768, immediately joining the city's informal circle of intellectuals. He was a charter member of the Royal Society of Edinburgh, founded in 1783. Edinburgh had by then already gained its formidable reputation as a seat of medical learning and practice,[4] and many members of this early Scottish intellectual circle were drawn from the ranks of the medical profession.

If Johnson had his Boswell, Hutton, too, had his explicator, in the person of one John Playfair (1748–1819), whose *Illustrations of the Huttonian Theory of the Earth* appeared in 1802.[5] Hutton himself was not much of a writer, and his initial paper, read at two sessions of the fledgling Royal Society of Edinburgh in 1785, was eventually published in 1788 as *Theory of the Earth; or an Investigation of the Laws Observable in the Composition, Dissolution, and Restoration of Land upon the Globe.* There was little reaction, except by supporters of Abraham Gottlob Werner (1749–1817). Werner was chief architect of an earlier theory, *Neptunism,* whose "geognostic" notions were widely regarded, even in his day, as fraught far more with a priori dicta on the way things are than with careful observation of earthly materials. According to Playfair, it was not until Hutton's chief critic, the Irish chemist George Kirwan (a Wernerian), finally got under his skin that Hutton was motivated to complete his two-volume *Theory of the Earth, with Proofs and Illustrations,* published in 1795, just two years before his death.[6] Only some 400 to 500 copies were printed, and by all accounts this work was a tough read indeed. It remained for Playfair, aware of the book's stylistic impenetrability and smarting at the misrepresentations of Hutton's thought, to set the record straight. He did so with admirable clarity in *Illustrations.*

Hutton is remembered chiefly for his formulation of the notion of *uniformitarianism.* Just as Darwin's later conceptualization of natural selection came—however incompletely and inaccurately—to be summed up in the catch-phrase "survival of the fittest" (coined, not by Darwin, but rather by Herbert Spencer), uniformitarianism is conventionally equated with the phrase "the present is the key to the past," an expression coined by Charles Lyell (1797–1875), whose three-volume *Principles of Geology* (1830–1833) completed the Huttonian revolution. Indeed, using knowledge of the nature of present-day processes to interpret the past is the core meaning of uniformitarianism.

In one of his earlier papers, Stephen Jay Gould teased apart the several nuances of meaning forever attached to the term *uniformitarianism.*[7] Gould makes a very useful distinction between what he called *methodological* and *substantive* uniformitarianism. The former is a way of thinking, the latter an assertion about the way the world really is—an assertion that Gould pointed out is testable and by no means necessarily true.

Methodological uniformitarianism is the essence of Hutton's gift to history. Gould notes that it amounts to nothing more, or less, than inductive reasoning: We make an underlying ontological assumption that physical processes operating in the material universe remain the same, from the earliest appearance of particular classes of material furniture, up through the present moment, and continuing for as long as such classes of furniture continue to exist. Thus we observe gravity, and adduce general mathematical descriptions about the gravitational relations between physical objects. We assume these general rules have held ever since there were macroscopic physical entities—and will forever continue to do so.

Some physical laws contain further generalities about *rates* at which events occur. Thus acceleration under gravitational attraction involving two physical objects occurs in the same general manner, no matter what the initial distance or mass of those two objects. But many of the physical processes affecting the earth have variable rates. Consider solar radiation: The product of radioactive decay, the sun one day will run out of nuclear fuel, and the earth (if it survives the sun's red-giant phase) will eventually be dark and cold. Similarly, the earth–moon gravitational system, while eminently describable and predictable under gravitational law, is destined to change: The moon will eventually escape the earth's gravitational field. And the earth has been slowing down. It now spins 365 times per rotation around the sun, down some 35 spins from Devonian times just under 400 million years ago. This slowdown was predicted from theory and has actually been measured from the daily growth lines of fossil corals.

Thus it is one thing to assume that the earth was revolving around the sun and spinning on its axis 400 million years ago. It is quite another to assume that the planet was spinning at the same rate then as now. The former is a safe assumption; the latter—Gould's *substantive* uniformitarianism—is not. And it is one thing to assume that processes we see going on around us today were going on in the past, however it is quite another to assert—as many geologists, particularly after Lyell, once did—that characteristic rates of, say, deposition of sediments observed in a particular lake are characteristic of all lakes, anywhere, at all times.

The fallacy of such a perception would seem obvious, but in fact it was not to early enthusiasts of uniformitarianism. The distinction between methodological and substantive uniformitarianism was missed initially because uniformitarianism was viewed as the antithesis to *catastrophism*—a prominent rival theory in geology. Early (post-Lyelian) uniformitarians made the strong claim that we can interpret past events in terms of modern-day observable processes *acting at the same, generally slow, and incremental rates* we see in action today. This claim countered the charge that earth history has consisted of a series of abrupt, relatively catastrophic events interrupting longer periods of relative stability.

When Hutton was formulating his thoughts in the last two decades of the eighteenth century, the rival system, as already briefly noted, was associated primarily with the German Abraham Gottlob Werner. Werner was the chief architect of Neptunism, a theory in which he insisted that all the rocks of the earth's crust are either chemical precipitates or aggregations of sedimentary particles deposited on the floor of a primeval ocean that completely covered the planet. The whole earth, down to its very core, came from that ocean, though how the ocean could exist before the initial phases of the earth were precipitated was apparently never addressed by Werner or his followers.

Rivals were called *Plutonists*, because they believed that some rocks are the cooled, hardened results of fiery molten masses. But, as John Playfair pointed out, Hutton himself couldn't be pigeonholed: He saw some rocks as sedimentary deposits, others as precipitates, and still others formed from molten bodies. For example, Hutton interpreted the landmark Arthur's Seat (named for legendary King Arthur), adjacent to Holyrood Castle in Edinburgh, as a sill—a body of molten igneous material pushed between layers of sedimentary rock. Crucial to Hutton's analysis was the manifest baking of the sediments at the contact between the molten mass and the older sedimentary layers along which the mass was intruded.

Pigeonholed or not, Wernerians detested Hutton's work, for a while being the only ones to pay it much heed. But, as the century turned, another school arose. This one was based on the very solid work of Baron Georges Cuvier (1769–1832), an admirer of Werner, whose major opus *Discours sur les Révolutions de la Surface du Globe* appeared in 1812. By no means given to apriorisms on the nature of things, Cuvier, known to history as the "father" of comparative anatomy, was deeply imbued in the study of vertebrate fossils, chiefly from the environs of Paris. His *Ossemens Fossiles*, also published in 1812—a monumental monograph still consulted by modern paleontologists—is massive testimony to his dedication to painstaking empirical research.[8]

It was Cuvier's renown for empirical research that gained him, and his notions of catastrophe, such a prominent following. Cuvier's basic thesis held that no fewer than 33 major upheavals had occurred during life's history—each one entirely wiping out the fauna of the time, and each followed by a separate act of creation to produce the new denizens of earth. There was little intrinsic to the older Wernerian views that called explicitly for such revolutions. But the fact that Cuvier saw his revolutions occurring in concert with marked changes in the rocks containing his fossils, and that these geological layers were arranged like concentric spheres within spheres, served as a sort of modernized version of the older Neptunism.

No one appreciated Cuvier more than the Reverend William Buckland (1784–1856), an Oxford clergyman and Reader in Geology who, especially in the decade of the 1820s, was England's dominant geological thinker. Buckland's main goal was to reconcile geological observation with the cosmological accounts of creation in the biblical book of

Genesis. The Noachian deluge, of course, was *the* quintessential catastrophe. And modern creationists are still fond of citing Buckland on the sedimentological and stratigraphic evidence for a worldwide flood.

It was against this entrenched—and, at least in the case of Cuvier, empirically respectable—backdrop that the likes of Charles Lyell, and Hutton and Playfair before him, struggled. Lyell, originally trained as a lawyer, became the one geologist who was able to tip the scales finally in terms of reasoned, empirical research in geology. He did so through his three-volume *Principles of Geology*—volume 1 of which Charles Darwin took with him on the *Beagle*. (Volume 2 caught up with Darwin during the five-year voyage.)

To prevail, Lyell had to insist not only that geological phenomena could be rationally investigated through a combination of careful observation and knowledge of earthly processes but also that these processes act, in a sense, timelessly. They act, for the most part, very slowly, and the large-scale effects that we see in the geological record, up to and including the building of mountains, are the outcome of slow, steady incremental effects over inordinately long periods of time.

This was the origin of the cardinal postulate known as *gradualism*: the notion that processes molding the history of both the physical earth *and* its living inhabitants proceed at a generally slow and *regular* pace. Gradualism is closely associated with another postulate involving eternal, observable physical processes: methodological uniformitarianism. The latter has a tacit corollary to the central notion of gradualism: No additional processes *not* observable in the present underlie elements of earth and evolutionary history. Melded with methodological uniformitarianism, gradualism accounted for many of the early triumphs of geology and biology. But as an a priori dictum on characteristic rates and scales of physical and biological phenomena, gradualism has also slowed the pace of discovery of larger-scale entities, events, and processes. These phenomena are immanent in the present world but difficult to perceive simply because of scalar considerations, especially large-scale and long-term ones.

As a quick example from modern disputes in evolutionary biology, consider the process of speciation—the derivation of two or more descendant species from an ancestral species. Speciation is commonly regarded as requiring, on average, from several hundred to several thousand years to complete. To an experimental biologist, the process is hopelessly slow. After all, no utterly convincing case of true speciation (that is, involving sexually reproducing organisms) has as yet emanated from a genetics lab. On the other hand, a paleontologist, especially one like myself, who works routinely with fossils hundreds of millions of years old, ordinarily finds it extremely difficult to measure time in such *small* increments as centuries and millennia. To us, speciation seems almost blindingly quick, especially when contrasted with much longer periods (millions of years, often) that species appear to persist relatively unchanged.[9]

Thus gradualism holds that processes observable today are both necessary *and sufficient* to explain the past. As humans, we are limited in our observations to the framework of tens, perhaps now hundreds, of years. Science, by probing the ultrasmall domains, has uncovered ephemeral entities and startlingly fast processes unimagined by our forerunners of the nineteenth century. Yet, as I remarked in Chapter 1, a comparable understanding of large-scale events and processes has been much slower in coming in both geology and biology.

I have just alluded to species and speciation as large-scale biological systems. For an additional example, consider the existence of large-scale "plates" dividing up the earth's crust. These are delineated for the most part by zones of active volcanism and earthquakes. Yet no special attention was once given to such patterns as the "ring of fire" marking the boundaries of continents and the Pacific Ocean. None, that is, until it was realized that such seismic zones mark the very boundaries where plates are meeting, rubbing past one another, diving under one another, or (in midocean ridges) diverging from one another. We now have a theory of *plate tectonics* that sees new crust generated in proportion to the destruction of the old. Such a theory could not even exist were the large-scale divisions of the earth's crust themselves not recognized. We needed to see the structure of the crust before we could generate a dynamic theory of it.

Much the same pertains to time itself. As we shall see later in this chapter, the true magnitude of geological time began to emerge no later than the mid-eighteenth century. But it is still somewhat mind-boggling, even to a grizzled paleontologist, that the earth is over 4½ billion years old, that life is at least 3.5 to 3.8 billion years old, and that it took an additional 2 billion years or so for complex, multicellular life to appear. Of course, deep time does not automatically imply the existence of larger-scale phenomena operating at rates not easily observed in processes going on around us right now. But the sheer magnitude of geological time raises issues of just how laboratory-observed rates of, say, genetic change really do mesh with thousands, millions, and billions of years.

Not that early geologists such as Lyell insisted rigidly that all phenomena of the geological past took place at the same slow pace. Lyell came relatively late in a long string of geologists who visited the rounded hills of the French province of Auvergne, so he was not the first to (correctly) interpret what he observed there as the product of volcanic activity. The main point of the Auvergne volcanoes was the death knell they rang on the old Neptunism and the triumph of applying present-day observable processes of volcanism to the interpretation of old geological structures. Yet it did not escape Lyell's notice that some processes, such as the spewing forth of lava, are characteristically more abrupt, accumulate faster, and are over sooner than, say, the more measured pace and slower accumulation of sediments in a marine basin.

The strength of uniformitarianism is clearly the commonsensical reliance of interpretation of the past strictly on what we think we have established about how the world works—what exists, and what things do to one another. Dobzhansky put this noble and undeniable notion extremely well when he said that, however "reluctantly," we must "put a sign of equality between the mechanisms of macro- and microevolution." By that he meant that we must interpret all large-scale events in evolutionary history as strictly the outcome of those evolutionary processes we have been able to establish as the result of experimentation and observation in the field and laboratory.[10] For example, large-scale events like the appearance and evolutionary history of major groups of organisms, (on the order of the cat family or flowering plants) must be interpreted in light of observable processes such as mutation, natural selection, and speciation.

The obvious weakness of uniformitarianism is the difficulty we have in apprehending phenomena beyond the normal domain of human sensory perception. If technological developments have allowed us to probe the ultrasmall domains, we have not been so fortunate in devising instrumentation to probe the nature of larger-scale entities—in either time or space. It is the implications of patterns in the historical record of life, the earth and the rest of the material universe that, ultimately, force us to abandon the rigid of apriority of uniformitarianism. We must be prepared to look beyond that which we can see right now around us— the easily observable entities and processes, and their characteristic rates. We routinely assume that these beings and events are wholly sufficient to formulate a complete explanation of natural phenomena. Often patterns come to the breaking point, literally forcing a new look at present-day phenomena. As is so clear with the story of the emergence of the study of plate tectonics, it is entities and events of the past, strung out over large spatiotemporal scales (in this case, large chunks of the earth's crust shifting over millions of years), that have compelled geologists to look again at the processes going on today.

Then, too, uniformitarianism and its commingled twin, gradualism, have become nearly synonymous with reductionism. The notion that what we can see around us today is both necessary and sufficient to explain the evolutionary histories of the earth and of life has become a rationale for probing no further. In some instances it has meant insisting that the data and theory of one particular narrow outlook or discipline are all that collectively we will ever need to reach full understanding. Examples in contemporary evolutionary biology come immediately to mind.[11] But consider as well the negative impact that Lyell's championing of uniformitarianism had on mid-nineteenth-century geology. Because catastrophism had been linked both with the earlier Neptunism *and* with ongoing attempts to reconcile the emerging facts of earth's history with the creation accounts of the book of Genesis, the careful, empirical work of the likes of Georges Cuvier were deeply eclipsed. Cuvier's work had made it abundantly clear that events of unusual magnitude and rate from time to time pepper the pages of earth (and life's) history. But that work was largely ignored for the bulk of the nineteenth and the first half of the twentieth centuries.

To a very great extent, the subsequent story of both historical geology and evolutionary biology has been one of trying to put back together what were seen as incompatibilities, at least from the late eighteenth through the mid-nineteenth centuries. The rational, scientific understanding of physical and biological history through the patient application of knowledge of the present-day natural world has had to be reconciled with early suggestions that many phenomena of the historical record lie outside the rote insistence that everything is the accumulated outcome of slow, steady incremental processes. Proponents of strict uniformitarianism continue to suggest that the rational contemplation of earthly and biological history is jeopardized by postulating the existence of large-scale entities and events that *are not the logical, projected outcome of "normal," everyday observable rates of wholly familiar processes.* In fact, the two are by no means incongruent, as no rational person suggests that the laws of nature were different in the past than they are now. After all, volcanoes have been active throughout earth history, and indeed exist on other bodies in the solar system. And organisms have always had genes. But, at the same time, patterns in the history of the earth and of life suggest that, almost routinely, these factors can work, and often do, in different characteristic combinations, rates, and scales. The results are wholly unpredictable from the simple extrapolation of commonly perceived events and processes now taking place around us.

So a connection had to be forged between uniformitarianism, gradualism, and reductionism: *extrapolationism,* the projection of commonly observed rates and processes as a prediction of what history *ought* to look like. When used to form a null hypothesis against which actual patterns in the history of the earth and life can be compared, such extrapolationism is incredibly valuable. For example, one assumption posits that species lineages evolved in a generally slow, steady pattern—a conception that Gould and I have labeled *phyletic gradualism*. This has been a prediction standard from the days of Darwin himself. Such an assumption contrasts strongly with observed patterns of stability and relatively rapid change encountered in the fossil record. Phyletic gradualism has always been an unexamined assumption of the characteristic tempo and mode of evolution. But as a null hypothesis—an extrapolated expectation of what the pattern of evolution *ought* to look like—that assumed pattern has turned out to be far from the typical historical trace left by the evolutionary process.

When a prediction actually matches an observed pattern, we can only claim consistency: Something else might be producing that same pattern. But we are way ahead if we can show a conflict between predicted and observed pattern. This is precisely what happened when extrapolated patterns of phyletic gradualism collided with empirical patterns of stasis and change within species lineages preserved in the rock record. Indeed, most advances in understanding in both historical geology and biology over the past 150 years have come from observations failing to agree with predicted patterns. Those predictions, for the most part, arose as small-scale processes extrapolated over larger-scale domains of space and time.

There is a naive form of extrapolationism, where the results are predicted but nobody bothers to check them against the patterns of history.[12] Results from this type of extrapolationism, where apriority reigns supreme, are antithetical to the growth of scientific understanding. There are plenty of examples of this less-than-helpful aspect of extrapolationism as well. But, first, consider some initial triumphs of extrapolationism: the emergence of the concepts of deep time and decipherable history, and the relationship between pattern and process in both forming and deciphering history, leading up to a coherent evolutionary theory of the earth, on the one hand, and the entirely commingled, yet simultaneously curiously detached, biological theory of evolution, on the other.

Deep Time, Evolution and History: Early Triumphs of Extrapolationism

Life evolved on earth,[13] and the oldest fossils yet discovered are in the oldest sedimentary rocks known—implying that life is intrinsic to the earth. Correlation between events in the evolution of life and earth history are very familiar to paleontologists. For example, we know that global cooling events caused massive ecological disruption, as well as extinction and evolution of new species. The history of our planet has profoundly affected the history of life; the reverse is true as well: life-forms have transformed the very face of the earth (the existence of an oxidizing atmosphere is due to the evolution of photosynthesis).

Why, then, did a full century separate the development and essential acceptance of a dynamic theory of the history of the earth, on the one hand, and of life on the other? Even more puzzling, why was it that the notion of *biological* evolution came so much earlier than an accepted evolutionary theory of the earth? For, even though a strict interpretation of the Genesis account of creation called for an earth created on the same day as the first living things,[14] we humans seem to have had an easier time contemplating the nature of objects farthest removed from the human condition per se. For example, early Greeks, Persians, Egyptians, Indians, Chinese, Incans, and Mayans number among the many cultures known to have made systematic observation of the heavens. In most instances, they did so before the Greek Eratosthenes demonstrated that the earth is spherical.[15] And certainly these cultures did so before any of them had made similarly systematic observations of the living world around them. Also, even after biological evolution had become accepted as *the* explanation for the history of life, some of its earliest proponents (including Alfred Russell Wallace!) denied that all aspects of the human condition were explicable under this rubric. It seems that what has often been claimed is true: that it is easier to be more dispassionate, hence conceivably more objective, about objects the less they directly impinge on the basics of human existence.

Despite this situation, we have had a coherent theory of biological evolution for more than a century before we have had an analogous, dynamic theory of the earth. (At least one that has pulled together all the disparate elements and most of the unexplained puzzles that geologists had been unearthing since the late eighteenth century). Why? How could this be?

The answer seems to lie in the relation between patterns, on the one hand, and the availability of plausible mechanisms to produce such patterns on the other. Bold, obvious patterns in both geology and biology seemed to cry out for a notion, not only of deep time, but of a regularity of historical pattern, of a history reflecting a seemingly endless string of events-consequent-on-events, of a temporally linear evolution—Such patterns cry out for explanation—processes understood from the world around us that might account for such stupendous change through time.

The simple answer, then—the one that holds most of the truth—is this: though there were patterns aplenty suggesting not only that the earth had continued to change through vast amounts of time, but had done so in a reasonably understandable, coherent way, geologists could not identify enough elements of the underlying process to allow them to make the final leap. This was not the case in biology. Darwin's and Wallace's notion of natural selection seemed to provide enough of a plausible mechanism to allow the scientific world to agree that the patterns that seemed to call for such a theory indeed meant that life had evolved.

That is the simple explanation. But the details involve the very same problems of scale in observing critical features that have haunted the history of both fields. When I say "coherent evolutionary theory of the earth," I refer, of course, to modern-day plate tectonics—the legitimate heir to the never-accepted concept of continental drift. But the *rudiments* of an evolutionary theory of the earth were in fact in place. They became more generally acceptable, beginning in the late eighteenth century, when early forerunner announcements of biological evolution (from Lamarck, Buffon, Erasmus Darwin, and so on) were falling on largely deaf ears (just as early twentieth-century calls for continental drift were destined to do.) For those geological evolutionary roots, and for the emergence of the concept of deep time, we return once again to James Hutton.

Although Hutton saw "no vestige of a beginning, no prospect for an end," he did see an endless cycle, where wind and water slowly ate away at the landscape. He saw rocks reduced to mineral grains, carried away to lake basins and eventually to the sea; he saw vast deposits of sediments, hardened into rock. To him, these deposits were obviously accumulations of these mineral grains that slowly, gradually filled the basins.

But Hutton saw something else, as well: He saw fresh, relatively unworn promontories, some made of hardened lava flows, some composed of thick piles of sediments. And there was Hutton's major *"aha!"* He saw that mountains, cliffs, and spires were often made of rocks that themselves were the accumulations of sediments *which had accumulated through the erosion of other mountains, cliffs, and spires.*

The concept is truly mind boggling! Mountains are themselves composed of the eroded remains of ancient mountains; continents are worn down, only to be rejuvenated. There was no other way to explain the interminable cycle that Hutton saw below his feet in Scotland. There must be some process raising the land, something to counterbalance the inexorable work of erosion that begins to gnaw at any piece of landscape. If there weren't, then the Genesis description of the initial stage of the

Earth "without form and void" would have to be true for all time.

Striking confirmation came when Hutton took a boat along the Scottish coast to Siccar Point—perhaps the most poignantly famous spot in the history of geological study. As shown in the accompanying figure, Hutton had seen the deformed and contorted beds of mudstone and sandstone, and he realized they were older than beds of the "Old Red Sandstone." Later workers realized that the older beds were Silurian in age—while the Old Red is Upper Devonian—a gap now understood to be on the order of tens of millions of years in the ages of these rocks.

Because Hutton naturally concurred with the earlier conclusions of the Danish physician Nicolaus Steno (also known as Niels Stensen, 1638–1686), he took his boat out to search for a specific pattern in these rocks. Steno had proposed two fundamental laws of sedimentation: The first law, "original horizontality," states that the layering of sediments is

James Hutton's famous unconformity at Siccar Point. Old Red Sandstone strata, gently inclined to the left, overlie nearly vertical strata of Silurian age — eloquent, if mute, testimony to powerful forces within the earth. In such patterns, Hutton saw "no vestige of a beginning, no prospect for an end" in the endless cycle of uplift and erosion in earth history.

originally horizontal, thus, tilted piles of sediments have been secondarily moved. The second, the "law of superposition" states that the sediments that get there first lie on the bottom of an accumulating pile (that is, sediments on the bottom of the pile are the oldest). Hutton knew that the bottom of the Old Red layers must lie in contact with the deformed older Silurian layers below. Most outcrops were either of one rock type or the other. But somewhere the contact had to be exposed. It should show, Hutton predicted, an angular contact between the contorted beds below, and the placidly flat-lying sediments above.

It was there, at Siccar Point, that Hutton found his contact. There, as he put it, he found "the ruins of an earlier world" overlain by the sediments that had eroded from the peaks whose very roots they now covered. Here lies "Hutton's unconformity," a "type specimen" of angular unconformity, a mute pattern that nonetheless cries out so eloquently for the existence of deep time and a dynamic earth.

The renowned British geologist Arthur Holmes (1890–1965), respected almost in spite of his being one of the few scientists of his era to accept the notion of continental drift, wrote in his monumental text *Principles of Physical Geology*[16] that Hutton's "supreme genius . . . lay in his demonstration that the earth is a thermally and dynamically active planet, internally as well as externally." Exactly so: For, to solar- and gravity-driven erosive powers of wind and water, Hutton also added "subterranean heat" to explain the uplift of rocks, the metamorphism of sedimentary rocks, the injection of granites and the outpourings of lava, and the tilting of strata along the sides of intrusive granitic bodies.

Hutton provided enough evidence that internal and external forces continually act to shape the earth for at least the rudiments of a dynamic theory of the earth to be established. But what Hutton described, his endless cycle, led, if anything, *away* from a sense of "evolution"—at least in the commonly implied sense of linear change through time. For Hutton was struck, first and foremost, by the *cyclicity* of these very general processes.

He was also struck by the enormity of time that seemed to be involved. Even given the rudimentary observations up through his time, it was well appreciated that muds and sands accumulate on seafloors very slowly. We now realize that submarine slumps leading especially to turbidity currents, which periodically race down continental slopes can deposit tens of meters of sediments in a few minutes' time. On the other hand, deep-sea deposits of the tiny shells of planktonic organisms build up at extremely slow rates at the opposite end of the spectrum. But, on average, the kinds of deposits of marine sands and muds that Hutton was considering accumulate at rates of a centimeter or so per year. And

Hutton was looking at really thick piles of sediment.

Everything in Hutton's cycle pointed to vast amounts of time—time many orders of magnitude greater than the conventionally allotted biblical time. Time in the millions, rather than thousands, of years. Hutton gave us this time, along with his essentially correct vision of an externally and internally dynamic earth, and his notion of continual cycling. He gave us the tools to decipher earth history, and he did it by realizing that patterns in the rocks have direct implications for the existence of processes forever at work. Patterns, such as his famous angular unconformity at Siccar Point, imply the former existence of mountains. They *had* to have been there. And he taught us to look as well to the present, to erosive forces of wind and water, and also to earthquakes and volcanism. Such processes provide testimony to the external and especially internal dynamic forces that we might plausibly invoke to explain such marvels.

Thus is the positive use of extrapolationism: From all his experiences, but especially from the telling pattern at Siccar Point, Hutton knew the earth must be incomprehensibly old. By relying on pattern to govern the invocation of process, Hutton truly did establish the modern science of geology.

The Geological Timescale

Hutton's insights, like Darwin's just over a half century later, opened up entirely new vistas, new fields of endeavor. If the earth has had a history, and that history is written in the sedimentary, metamorphic, and igneous rocks lying exposed about the earth's surface, then it becomes an obvious task to go out and piece that history together. A natural conclusion, and only common sense. But therein has lain the seeds of divorce between the details of the histories of both earth and life from the processes underlying those histories. For it falls to the historian to piece together segments of history. And it falls to the "functional" biologist and "physical" geologist to focus on the processes that produce that history. Soon, gone is the very connection that Hutton forged. Once the *fact* of deep time is established, the very sorts of recurrent patterns that forced that recognition in the first place tend to be brushed aside. At least such is the story in evolutionary biology. A somewhat brighter story unfolds in the modern history of plate tectonics, possibly because the facts of geological history were far more completely worked out by the 1960s when this dynamic evolutionary theory of the earth belatedly became established.

In any case, post-Huttonian geologists almost instantly began the task of deciphering earth history. The undertaking was very much like working out a vast puzzle. The earliest work focused on the conspicuous layers of sedimentary record containing, as a rule, a rich and varied fauna and flora of fossil organisms. And once again, the clues to assembling the puzzle lay in patterns in the rocks, especially patterns of occurrence of fossils in those sedimentary strata.

Baron Georges Cuvier played an early role in deciphering this puzzle. In addition to his visions of 33 successive cycles of extinction and re-creation of life through geological time, Cuvier got his hands dirty in the field as well as in the dissection lab. He produced yet another milestone in the history of science, with his colleague Alexandre Brongniart (1770–1847): one of the very first geological maps.

Paris, focal point of Cuvier and Brongniart's mapping, is built near the center of a shallow bowl of sediments—chiefly chalks and other forms of limestone. Though the rocks at each outcrop appear to lie absolutely flat, they are all inclined very slightly toward the center of this bowl—toward Paris, that is—so that approaching the city from any direction is to move slowly up through Cretaceous chalks into the fossiliferous beds of the younger Eocene. The entrenched Seine snakes westward from Paris toward the English Channel, traversing progressively older layers of Cretaceous chalk. Indeed, Cuvier got many of his Eocene fossil mammal specimens for his *Ossemens Fossiles* monograph from Montmartre, within the confines of Paris itself.

Thus, because sedimentary strata are often slightly tilted by earth movements, rocks of different ages intersect the surface in different places. But they do so in regular patterns. And different elevations expose younger (higher) and older (lower) rocks from place to place. This was the great discovery of Cuvier and Brongniart in France, mirrored, extended, and developed further in parallel work by William Smith (1769–1839) in England. (Smith was also known as "Strata"—a horrible sobriquet.) As shown in the figure on the next page, by tracing out zones where the same beds of rock are exposed ("crop out"), a topographic map can be converted into a geological map, one whose regular patterns (usually depicted in brightly contrasting colors) reveal the relative ages, and thus something of the geological history, of a region at a glance.

Smith worked as a surveyor in the early decades of the nineteenth century, when the great canal systems that played such an integral part of Britain's industrial revolution were being dug. Smith trudged from hill to hill, setting up his transit and mapping the course of the canals. In doing so he noted the regularity with which certain kinds of fossils invariably occurred. Much of Smith's work was in the English countryside north of London—where rocks that later came to be called Jurassic (from the Jura mountains of France and Switzerland) form the underpinnings of the

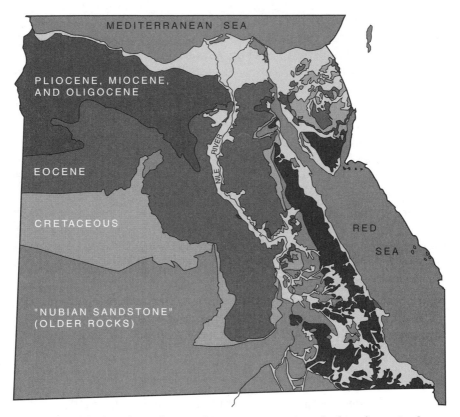

In this simplified geological map of Egypt, contrasting shades of gray in the Sahara west of the Nile reveal a simple north–south pattern of progressively older strata outcropping, from the Pliocene, Miocene, and Oligocene in the north to the older "Nubian sandstone" to the south.

landscape. As he climbed those hills, Smith saw oysters, snails, and other mollusks—particularly ammonites, the often beautifully logarithmically coiled external shells of animals closely related to squids and octopi. The ammonites are extinct now, having disappeared forever in the great extinction event that also took out the dinosaurs and other groups, at the end of the Cretaceous Period of geological time, some 65 million years ago. But they flourished for an incredibly long period of time, beginning their sojourn on earth in the Silurian Period of the Paleozoic Era, over 400 million years ago.

Just as Cuvier and Brongniart were noticing with their Cretaceous and Eocene rocks and fossils, Smith saw that recognizably different beds of rocks, often with their own particular suite of ammonoids and other fossils, monotonously occurred in the same sequence, over and over again. Smith understood that the layers were once continuous; they had formed as vast expanses of sea bottom accumulated layer upon layer as the geological ages rolled. Erosion subsequently separated these once-continuous layers, their remnants now confined to hillside outcrops. But Smith could match them up just by looking at the rocks—and especially at the fossils typically found in each layer.

However, as his peregrinations took him further afield, Smith found that layers on top of a hillside in one region became the bottom layers of hills miles away. And layers from the latter hills would be topped by still-younger layers missing from the first outcrops he visited. And so the picture-puzzle piecing together by recognizing the "same" layers miles away could be extended: As ancient tiltings in the earth's crust brought progressively older, or younger, rocks to the surface, their general relations to one another could be worked out. As shown in the accompanying diagram, the sequence of fossils, in general, is invariant, repeated over and over again. Use of the fossils as a guide enabled a general picture of a regional *geological column* to be pieced together—a vertical rendering of the snaky bands of strata depicted on the geological map.

There was lots of pattern, but no real process, here, since Cuvier, Smith, and those who quickly followed their lead were working in pre-evolutionary days. In Cuvier's case, perhaps it is more accurate to say he *did* have a sense of process underlying his empirical patterns: He saw stability of faunas and floras over periods of time, only to be at last disrupted by a destructive wave of extinction, followed by a new episode of creation by a (supernatural) Creator. But such interpretation amounted to little more than an underlying assumption. Smith, in contrast, was more typical of what later became the discipline of stratigraphy. He apparently cared little why there was order to the occurrence of his fossils, why certain fossils invariably occurred together, always above certain others—and below still others. To Smith, it only mattered that it was so.

After Darwin, of course, it became clear *why* there should be a change in the composition of life as represented by its fossilized remains in successive series of rock strata. Some geologists contemporary with Darwin (chief among them Charles Lyell[17]) did resist the idea of evolution when first convincingly articulated by Darwin in 1859. But stratigraphers soon thereafter came to realize that evolution, together with extinction, were the twin processes that created change through geological time in the characteristic composition of fossil floras and faunas.

Smith's exploration of Jurassic-aged marine rocks was a particularly happy coincidence for the fledgling field of stratigraphy. Ammonoids, like their squid relatives, were active swimmers, and the same species can often be found over wide areas. Though some groups remained

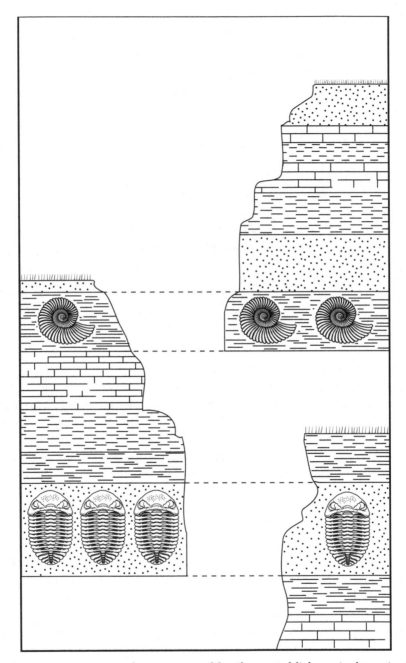

Geologists use patterns of occurrence of fossils to establish equivalency in the age of isolated bodies of rock. This diagram shows three separate outcrops, two of which share a Devonian trilobite species, and two of which share a Mesozoic ammonite species. In general, many species are used to establish such "correlation" between rock strata.

anatomically very conservative, producing few new species and changing little, other ammonoid groups typically evolved very quickly, with new species arising and, just as importantly, becoming extinct, at relatively rapid rates. Where most other marine invertebrates turn out to have average lifespans in the 5- to 10-million-year range, many ammonoids lasted only 1 to 3 million years.

Relatively quick turnover and good geographic spread are a stratigrapher's delight. You can "correlate" (that is, demonstrate that strata are roughly equivalent in age) more widely and accurately with such fossils. The late nineteenth-century German stratigraphic paleontologist Albert Oppel (1831–1865) was a student of Jurassic ammonoids. So was the Englishman W. J. Arkell (1904–1958), who developed and popularized Oppel's basic ideas of fossil zonation in the English-language geological literature. Oppel and Arkell both saw the fossil record of ammonites as forming a succession of "zones." Each zone had its own characteristic ammonites, many of which were unique to a single zone, with fewer species occurring in two or more successive zones. Oppel, rather like the much earlier Cuvier, interpreted the succession of faunas revealed by his empirical studies as the product of rapid bursts of evolution, followed by extinction. In essence, his was a Darwin-inspired process interpretation of the same kinds of patterns Cuvier saw in supernatural creationist terms.

However, there is a seldom-noted fact in considering the addition of evolutionary process theory to the practices of stratigraphic correlation using fossils. That is that all such early students—Cuvier, Brongniart, Smith, Murchison, plus the American James Hall (1811–1898), the Czech Joachim Barrande (1799–1883), and many other nineteenth-century paleontologists, right on up through Oppel and Arkell to modern times—all saw the stratigraphic persistence of species in most cases completely through the "zone" in which they occur. A zone in Jurassic time averages out to about 1 million years; in the Devonian, it is more like 5 to 7 million years.[18] The system is inherently empirical, and it is based on the persistence, through much or all of the duration of the zone, of the various different fossil species. As we shall see, this empirical situation is hardly what Darwin had in mind. The post-Darwinian wedding of stratigraphic empirical pattern with evolutionary process was *not*, in the main, a wedding of straightforward Darwinism with patterns in the history of life.

But pre- as well as post-Darwinian stratigraphers scarcely cared for the details of life's history except insofar as they shed light on the details of the *earth's* history. The 1820s to 1870s were the decades of the founding of the geological timescale, the one still in use today. When Darwin ventured out on one of many of Sedgwick's summer trips to Wales, he was taking part in the early stages of construction of the geological column for that part of the world.

Sedgwick was working on a series of layered rocks he called the Cambrian. (Cambria was the Roman name of Wales.) His friend and colleague Roderick Impy Murchison, who, in the early 1830s, accompanied Sedgwick to the field, was working a bit farther to the north. Murchison was working down from the Old Red sandstone (of Huttonian fame, now known to be Devonian in age). Blessed with richly fossiliferous rocks, Murchison named his sequence the Silurian (after the Silures, the Roman name for an indigenous tribe of the region). Sedgwick, on the other hand, seemed to be working from the bottom up; his oldest rocks, now known to be Lower Cambrian, were devoid of fossils.

Trouble erupted when, in the 1840s, the two men realized that the upper part of Sedgwick's Cambrian System corresponded to the *lower* part of Murchison's Silurian System. Tempers flared when neither man would back down, each drawing support from other gentlemen geologists, including fellow members of the Geological Society of London. If it was true that Murchison in the main attracted more favorable opinion,[19] the dispute festered nonetheless until both men were dead. The Solomonic Charles Lapworth split the difference, naming the Ordovician System (the Ordovices being yet another primordial tribe of the region) to recognize, formally, that Sedgwick's Upper Cambrian and Murchison's Lower Silurian were in reality one and the same.

But "one and the same" what? What are the Cambrian, Ordovician, and Silurian? What are the Devonian (named for the English region of Devon), the Carboniferous (the Mississippian and Pennsylvanian of North America), and the Permian? What are the Triassic, Jurassic, and Cretaceous (after *Kreta*, Greek for "chalk")? What are the Tertiary and Quaternary? And what, for that matter, are the Paleozoic (which includes the Cambrian through the Permian), the Mesozoic (Triassic, Jurassic and Cretaceous), and Cenozoic (Tertiary and Quaternary)?

All of these names correspond to bodies of rocks, thick sequences of layered sedimentary rock, first identified in a particular region. For example *Perm*, from which *Permian* is derived, is a Russian city in the Ural Mountains. Most of the names in the preceding paragraph label *systems*, which are bodies of rock corresponding to segments of geological time, or *periods*. And each such segment of time has the same name as the corresponding system. (This practice is convenient, if sometimes a bit confusing.) Thus, the Ordovician Period is that chunk of geological time between 510 and 440 million years ago, the span of time recorded in the original analyzed body of rocks in Wales bearing that same name.

James Hall and other early American geologists, working under the aegis of the New York State Geological Survey, were so impressed with the beautiful sequence of fossils in some of their rocks south of the Mohawk Valley and down into the southern tier of the state that they proposed, at first, to designate their findings the *New York System*, or the *New Yorkian*. They did so until they realized that New Yorkian corresponded exactly with the earlier-named British Devonian, although it was more complete and had, for the most part, more and better quality fossils. Priority rules, however, as it did a century later when some geologists proposed scrapping Silurian for Gotlandian: The rock and fossil sequence is unarguably far more complete on the Swedish Island of Gotland in the Baltic than it is in western England. These original places provided the hard evidence, the rocks and fossils, where the existence of a particular segment of geological time was first recognized. And, because the task of deciphering the chronological sequence of the pages of earth (and life's) history remains to a great degree a matter of determining "what is older than what," universal standards conforming to original definition must take precedence over later discoveries of more complete records.

Thus the geological timescale shown on the following page: a hierarchical array of divisions of geological time. As we shall see in the next chapter, biological classification reflects evolutionary history—and is based on the very same patterns that led, finally, to the acceptance of the very notion of evolution. So, too, with the geological timescale. Though modern creationists love to claim that the timescale is the work of pro-evolutionary zealots, nothing could be further from the truth. (Recall that Sedgwick, Darwin's mentor, was a clergyman and staunch opponent of evolution throughout his life). To the contrary, the divisions of geological time are *empirical*, reflecting for the most part the state of life through time. The last 540 or so million years are the Phanerozoic Era, the time of visible life. Before that we have the Archean (also known as the Azoic) and the Proterozoic, when bacteria were all there was to life on earth.

The Phanerozoic is in turn divided into the Paleozoic, Mesozoic, and Cenozoic Eras ("Ancient," "Middle," and "Recent" life), again reflecting major divisions in the history of life. We now know that two of the five major mass extinctions of the geological past transformed the face of life. And evolution essentially rebuilt all-but-lost ecosystems to form these three fundamental divisions. The divisions of each of these eras, the geological "periods," also are primarily based on objective patterns of extinction and evolution. So are even the finer subdivisions, such as Oppel's "zones."

ERAS	PERIODS	EPOCHS	years ago
CENOZOIC	Quaternary	Recent	
		Pleistocene	◄1.65 million
	Tertiary	Pliocene	
		Miocene	
		Oligocene	
		Eocene	
		Paleocene	◄ 65 million
MESOZOIC	Cretaceous	Carboniferous (Pennsylvanian + Mississippian)	
	Jurassic		◄ 208 million
	Triassic		◄ 245 million
PALEOZOIC	Permian		
	Pennsylvanian		
	Mississippian		
	Devonian		◄ 367 million
	Silurian		
	Ordovician		◄ 440 million
	Cambrian		◄ 540 million
PRE-CAMBRIAN			
			◄ 4.55 billion

A simplified chart of geological time, which emphasizes the most recent 540 million years. Here, the nested, hierarchical structure of the divisions of geological time stand out: epochs are parts of periods, which in turn are divisions of geological eras. Important absolute dates mentioned throughout the text are noted on the right side of the chart.

Thus was deep time delineated, classified, and charted. The process was nearly completed by the 1870s, though refinements of correlation and the much more recent calibration of actual ages continue. The early work was on the rocks at the upper end of geological time, now known to be just over the last half billion years of a history that is over 4½ billion years. This last half billion years was the time when life was rich, varied, abundant—and megascopic. It left an easily found fossil record that the earliest geologists could use to sort out and correlate their layered sequences.

How deep is deep time? It was as abundantly clear to the likes of Smith, Sedgwick, and Murchison as it had been to the earlier James Hutton that the thin layers of shale, corresponding to the gentle year-by-year accumulation of clay particles, implied an earth vastly older than biblical accounts would suggest.

By the mid-nineteenth century, the fellows of the Geological Society of London were comfortable with the thought that the earth must be millions of years old. But how many millions? Few could hazard a guess, though estimates seemed to center on a few tens of millions at the very least.

Enter, interestingly enough, Charles Darwin, and perhaps the last great contribution made from the still-new extrapolationism born of gradualism and uniformitarianism. Darwin's contribution to the apprehension of the true magnitude of deep time is enormous, and it is a contribution for which he rarely receives the credit he deserves. For no one did as much to establish the correct order of magnitude of geological time in preradioactive chemistry days than did Darwin.

Recall that Darwin, to the extent that he had received any formal scientific training at all, was more of a geologist than a botanist, zoologist or paleontologist. All of the fields fell under the rubric of "naturalist," which was his position on the epic five-year voyage around the world on HMS *Beagle* in the years 1831 to 1836. And he was to go on to make important geological contributions in his own right, most dramatically in his theory of coral reef formation published in 1842.

In his *Origin*, Darwin devotes two chapters to geological themes. Chapter 10, in keeping with the spirit of most of the rest of the book, is a lively invocation of patterns—in this instance, of organic forms in the fossil record—that Darwin invokes to establish the plausibility of the very idea of evolution. But the preceding chapter is a departure. Entitled "On the Imperfections of the Geological Record," Darwin is at pains to explain why the fossil record does *not* exactly fit his picture and thus suit his needs.

Darwin pins the blame on the record itself: The patterns are apparent, but they are not to be trusted. But the chapter is not to be ignored even if it was written to explain (away) the aching discrepancy between predicted pattern (of slow, steady gradual change) and actual experience (great stability of fossil species). For in challenging the essential integrity of the preserved patterns of the fossil record, Darwin managed to found an entirely new subdiscipline of paleontology, one which was not to blossom fully until the mid-twentieth century: so-called taphonomy.

In Chapter 9 of the *Origin*, Darwin says that the record of ancient life is skimpy in the extreme. Many vagaries of chance come into play between an organism's death, burial, preservation (or not) over geological eons, eventual exhumation, and, not least, showing up in some qualified observer's laboratory for considered analysis. He was, of course, right in saying all that. Thus, a new study had to come into being, one which examines how organisms become fossils in the first place: the aforementioned taphonomy. But, as we will later see, Darwin was quite wrong to suggest that these "imperfections" of the fossil record were sufficient to explain away the discrepancy between his predicted patterns and what paleontologists see routinely in the fossil record.

Yet what he had to say about the imperfections of the fossil record were true enough. And one aspect of his argument, ensconced in Chapter 9 of the *Origin*, was really all about geological time. Because the true depth of geological time had not as yet been established, *lack* of true deep time was to Darwin one of the "imperfections of the geological record." Yet, to Darwin, geological time was much deeper than most of his colleagues had ever proposed.

And that, almost ironically, is yet another of the myriad gifts to our understanding of the material universe we can credit to Charles Darwin. For Darwin *needed*—or felt he needed, at least—great chunks of time for his picture of almost ineffably slow evolution to work to produce the great diversity of life in all its myriad phyla, its great diversity of at least 10 million species now living on earth. In one particularly striking passage (on the "denudation of the Weald"), Darwin waxes eloquent on the slow pace of erosion shaping the landscape. He concludes that it must have taken 306,662,400 years, "or say three hundred million years," to remove the overburden from North Downs, assuming erosion removed 1 inch per century, and the initial cliff stood 500 feet above the present plain.

Well, the Weald is underlain by Upper Cretaceous chalks, the last of which formed around 70 million years ago. So Darwin's imagery was true hyperbole: He overshot the mark by a lot. But it is the order of magnitude of the estimate that is arresting. It was still a radical suggestion, in 1859, that anything connected to earth history could possibly be 300 million years old. Indeed, as late as 1898 the chemist John Joly (1857–1933) calculated that the entire age of the earth could be no more than 250 million years, based on estimates of the rate of erosion, and accumulation in oceanic waters, of the element sodium. Seawater is salty, its saltiness a reflection of the slow accumulation of sodium over time. Joly erred in not taking into account the sodium cycle. So he greatly underestimated both the rate of production of sodium ions through erosion and the amount of sodium precipitated in the vast salt deposits found on every continent.

Indeed, it took Ernest Rutherford's suggestion, in 1905, to apply Becquerel's and Curie's discovery of radioactivity to establish a truly accurate means of estimating the age of the earth. Finally there came available a means of dating rocks in terms of years, not just a relative system of What is older than what? By and large, it is only in igneous rocks where the ratio of "parent" to "daughter" isotopes can be measured and, with the rate of decay experimentally known, an age of formation of the rocks can be calculated. A complex web of different laboratories analyzing both the same and different decay pathways has led to an abundantly cross-checked and annotated system of "absolute" dates for all parts of the geological timescale. For example, granites and lavas associated with Lower Devonian fossiliferous beds consistently yield dates in the 390 to 395 million year range, but the Lower Devonian always emerges as earlier than the Middle Devonian and younger than the Upper Silurian (thank goodness). And rocks considered Lower Devonian the world over consistently come in at the same age, within margin of error.

So we have ample corroboration of the old, painstaking puzzle-piecing approach to assembling the geological timescale. We also know that Darwin was more nearly correct than many of his more purely geologically minded colleagues in estimating the order of magnitude of geological time. His estimates came more out of necessity—a by-product of his extrapolation of the pace of evolution—than ad hoc estimation of the rate of erosion of the Weald surface. But his grasp of the depth of geologic time was perhaps the last salubrious contribution of such gradualistic extrapolationism.

And by the end of the nineteenth century, process had been all but divorced from pattern in both historical geology and paleontology. Geologists and paleontologists for the most part had forgotten Hutton's lead, that patterns in the rocks really have much to tell us of the way things must be working in the natural world. Such was the main reason why continental drift/plate tectonics took so long to be developed and universally accepted. And it slowed the pace of understanding within evolutionary biology itself, which was to be born in full with the work of Charles Robert Darwin.

CHAPTER 3

Time, Energy, and Biological Evolution

Consider, once again the definition attributed (possibly apocryphally) to Ernst Mach: Science is the study of entities of the material universe and the interactions between them. Conventionally, science deals with particles—of whatever size, from subatomic "particles" to entire species or even ecosystems. By this definition, James Hutton was doing real science. Recall that Hutton realized that heat (energy) flowing from the earthly interior is the force driving uplift, and that solar-driven winds, rains, and gravity wear them down again; the whole cycle is repeated ad infinitum. The "things"—particles—in Hutton's conceptual universe were literally mundane entities: mountains, lava dikes and sills, and unconformities such as Siccar Point.

Because Hutton's notions were so relentlessly cyclical, no one thinks of his achievements as having established an "evolutionary theory" of the earth. Indeed, his star intellectual descendant, Charles Lyell, was such an adamant proponent of the earth's history as "steady-state," with no net change (itself a concept closely allied to cyclicity), that he staunchly opposed Darwin's attempt to legitimize the very notion of organic evolution. Lyell even doubted that any species had ever become truly extinct. Such a position is diametrically opposed to Cuvier's notions of catastrophic extinction and subsequent (divine) creation.

In contrast to the cyclicity of Hutton's views, and to Lyell's sense of earth history as steady-state, the term *evolution* has generally implied a sense of direction—of going from state A to state B in some more-or-less directional manner. Indeed, in French, the word *evolution* originally referred to embryological development, while, similarly, the German word *Entwicklung* took on the meaning "evolution" after a long-standing previous usage as "development." And in biology, then as now, *development* refers to the transformation of fertilized egg, through larval or other intermediate stages, to full-fledged adult: a development from a single cell to a complex multicellular organism. Embryological development always implies a directional change, a transformation, even an unfolding, through time.

So Hutton had a sense of both deep time and the physical forces continually shaping and reshaping the earth. But because his concept of earth history explicitly lacked a sense of direction, he is generally not regarded as providing a truly evolutionary theory of the earth. His contemporaries in what were to become the life sciences—his counterparts in the biological realm of plants and animals—came up with a strikingly different sense of the nature of the living world than Hutton achieved for the macroscopic, inanimate world of the earth. The earliest biologists preceding Darwin also failed to establish a true and universally accepted evolutionary biology. But they nonetheless perceived fundamental patterns that later proved to be the underpinnings of two central ideas: that life has had a long history, and that all organisms presently living on earth are descended from a single common ancestor living in the dim recesses of earliest geological time. What Hutton's biologically minded contemporaries failed to accomplish—even those like Jean Baptiste, Comte de Lamarck (1744–1829)—was to specify exactly how the material world, not so much its particles, but its *forces*, actually works.

Linking up the world of evolutionary history clearly and unambiguously with the world of energy flow has proven difficult. Indeed the full connection has yet to be specified unambiguously and categorically—a theme that we shall pursue vigorously throughout the remainder of this narrative. For the moment, we turn to patterns revealed through the efforts of Hutton's biological contemporaries. Such patterns hold the seeds of directionality and other quintessential evolutionary signals. Hutton's contemporaries got closer to an evolutionary perspective than Hutton did, but Hutton had the advantage in having no difficulty discerning the rudiments of the physical forces underlying his chosen historical phenomena. Biologists contemporary with Hutton, for the most part, mostly stuck with God as the essential Creator of the very patterns their work revealed.

Scala Naturae and Natural Kinds

To this day no one knows how many species, or discrete reproductive communities, of plants, animals, fungi, and microbes currently exist. Some 1.75 million have been named and classified formally in the scientific literature. But estimates vary widely on how many there are as yet to be discovered or simply studied. There may be as many as 100 million, though the more conservative estimates of 10 million to 13 million, coming from ecologist Robert May and his colleagues, seem to be gaining a general acceptance. It is true that human transformation of the surface of the earth has plunged the biota into what is often called the "sixth" mass extinction,[1] meaning that there are demonstrably fewer species in existence right now than there were in the seventeenth and eighteenth centuries when comparative biology was in its infancy. Nonetheless present-day species diversity is still at approximately the same order of magnitude as it was in the days of the Swede Carolus Linnaeus (1707–1778), the Frenchmen G. L. L. Buffon (1707–1788), and Lamarck and the other pre-Darwinian students of biological diversity.

However many millions of species there may be, they are in aggregate a bewildering array. As Dobzhansky remarked at the outset of his 1937 monograph, *Genetics and the Origin of Species*, the anatomical diversity embodied in the panoply of life constitutes the original "problem" for the very notion of evolution to explain. That very notion of evolution is the postulate that all organisms on earth are descended from a single common ancestor that lived in the remoter sectors of geological time. Yet the notion itself had to be established before an organizing principle, such as natural selection, could be brought to bear as an explanation for the amazing anatomical diversity of life. And crucial to the recognition that life had in fact evolved was the recognition of some order in the otherwise chaotic variation that the organic world presents to the human eye and mind.

Though there remain occasional observers who claim that there is no inherent order, no internal structure, to biological diversity, it was clear even to the ancient Greeks that two very general sorts of patterns do lend such order to the biological world.[2] These two patterns are distinct, yet interrelated in complex ways. In fact, they may fairly be said to constitute two different aspects, or ways of looking at, the same thing: what we now understand to be patterns of genealogical descent linking up absolutely all organisms, past and present, living on earth.

The two patterns are, first, the *scala naturae*, or "great chain of being." Here, elements of biological diversity are ranked from simplest to more complex. Or, in a characterization more faithful to the earliest usage, they are ranked along a gradient from least to most perfect. For want of a better term, I'll call the second sort of biological pattern *natural kinds*—the recognition that resemblances among species are partitioned into groups, some more obvious than others, and that such groups may contain other, smaller, constituent groups.

    As an example that reveals both the natures of, as well as something
of the relationship between, these two sorts of patterns, consider
Lamarck's term *invertebrates*. Aristotle had divided the animal world into
"bloodless" and "blooded" creatures, which was a distinction Lamarck essentially renamed as "invertebrates" and "vertebrates," respectively.
Vertebrates are all those animals with a bony or cartilaginous backbone;
invertebrates are all those animals that lack such a structure. The pattern uppermost in Lamarck's mind was always the *scala naturae*, the gra-
dation from lower to higher. And just as Aristotle proclaimed the
"blooded" animals superior to the "bloodless," Lamarck saw invertebrates
to be subordinate in the sense of "less perfect" with respect to the higher
animals, the vertebrates.

    It was precisely this lower-to-higher sequence that Lamarck thought
implied direct connection, an evolutionary sequence in geological time,
from lower, more simpler forms on up through higher beings.[3] And there
is no doubt that there is a general scale of progression that suggests continuity within the animal kingdom, in refined but by no means unrecog-
nizable form, as first seen by Aristotle, Linnaeus, and Lamarck. The
lowliest of the 36 or so phyla, or major subdivisions, of the animal kingdom,[4] the phylum Placozoa, is barely distinguishable from a colonial, or
multicelled, microbe. Next come the sponges, with two or three sets of
different kinds of cells; however, they lack true tissues (interconnected
masses of similar cells), let alone true organs (generally formed from two
or more sets of tissues). Next along the invertebrate scale of the great
chain of life come the cnidarians—corals, jellyfish, and their kin—
which have true tissues but lack true organ systems. Flatworms and other
acoels (acoels lack true body cavities) come next; but, in contrast with
their "lower" invertebrate kin, they have true organs like nephridia (ex-
cretory organs), eyes, and a digestive track.

    The "higher" invertebrates all have true body cavities. Here we
reach a schism. There are two sorts of body cavities, which, together
with other anatomical, embryological, and biochemical information,
serve to delineate two co-equal groups. On the one hand, there are the
protostomes, which include annelid worms, such as leeches and earthworms; mollusks, such as clams, snails, squid, and others; and arthropods,
such as insects, spiders, scorpions, and crustaceans. The countervailing
groups are the *deuterostomes*, which include echinoderms, such as starfish
and sea urchins; some primitive "chordate" vertebrate relatives; and the
vertebrates. The deuterostome/protostome dichotomy may upset the linear array of simple to complex. But the fact that we humans, who are
vertebrates, are part of the deuterostome array led most early zoologists,
including many well within the modern era, to trace the "main line" of

vertebrate evolutionary descent through the lower deuterostomes to our "exalted" selves. The protostomes were considered a side branch, however diverse and interesting they might be.

The *scala naturae* has been widely derided. Apparently as a holdover from medieval scholasticism, some early advocates of the concept sought unity between the smallest particles of inanimate matter on up through fungi, plants, and animals. (Later they included the microbial world forever revealed through the work especially of van Leeuwenhoek in the mid-1600s). Thus it must have been with a degree of ambivalence that Lamarck accepted his colleague Abraham Trembley's conclusion that corals and other sedentary and superficially plantlike animals are not true intermediates connecting the plant and animal worlds, but are in fact animals. And the branching nature, the *non*linearity, of the sequence must have been disturbing as well. Though Lamarck knew nothing of the distinction between protostomes and deuterostomes, which was established in the latter part of the nineteenth century, it was clear that the spectrum of living diversity could not be smoothly pigeonholed into a rigid, single straight line.

Yet it is obvious that there *is* a pattern there. But it is one of gradations, from relatively simple to relatively more complex, not in one line, but within many.[5] The *scala* is pattern, and it reveals structure within that bewildering array of life's diversity.

Now, suppose you, like Lamarck, submit some refined version of the *scala* of biological complexity as evidence that life has evolved. Over the geological eons, life has progressed though a continual process of ancestry and descent from relatively simple on up through progressively more complex beings. Because there are so many different species in the world today, and because they display that very spectrum of simple to complex, you would have to suppose two things: first, that evolution in fact does branch, and second, that it must, or else there would be no way to account for the survival of the more primitive after the appearance of the more advanced. In other words, evolution couldn't have happened only through the transformation of an ancestor into a descendant, with the ancestor becoming extinct in the process. Otherwise, there would be no diversity, no living spectrum of simple to complex on which to base our entire conjecture that life has evolved in the first place! The question now becomes: How convincing is the argument that survival of the spectrum, this pattern of simple to complex, *necessarily implies* that life has evolved?

And here is an important lesson, for patterns themselves may strongly hint at the truth of a proposition, but seldom do they admit to a single definitive interpretation. Perhaps the most important principle in modern science is this: *If a proposition about the material universe is true, then we ought to observe a specifiable set of consequences.* In other words, if the principle is true, we should be able to make predictions about what we might observe. So we ask, if evolution is "true," that is, if all organisms are in fact descended from a single common ancestor, would we necessarily expect to observe a pattern that our early forerunners would characterize as the *scala naturae*?

Here the answer is ambiguous. We would predict the temporal evidence, such as the relative age of fossils, to conform to a general simple-to-more-complex pattern in the history of life. And we would imagine that the original life-forms (however life is defined!) would be vastly more simple in structure than even the simplest of existing bacteria. No one would suppose that life started out with cheetahs and, through a long process, ended up as bacteria.

Long after Lamarck, we know enough about the fossil record to assess the above prediction—and find it confirmed, at least in general terms. For example, the earliest fossils (some 3.8 billion years old) are simple bacteria. Complex cells (microbes with cells essentially similar to those of our own bodies—known as eukaryotic life) appear perhaps as far back as 2 billion years ago. More complex, multicellular animal life appears perhaps 700 million years ago. Unfortunately for Lamarck and his invertebrates, the proliferation of everything from sponges to chordates appears to have happened in a whirlwind of diversification requiring no more than roughly 10 million years, starting perhaps 540 million years ago. (Score one for Charles Lyell!) True vertebrates appear somewhat later than all the rest. Within land-living, or *tetrapod*, vertebrates, amphibians (with their simpler reproductive systems) predate reptiles, the first of the amniotic-egg producers; reptiles, in turn, predate both birds and mammals, as predicted. As far as plants go, the sequence is even more definitive, with the spectrum of such relatively simple groups as horsetail ferns and mosses on up through the gymnosperms (evergreens) and angiosperms (flowering plants) conforming very well to their relative appearance in the geological record.

But there are flies in the ointment. The *scala naturae* is that component of diversity pattern that emphasizes linearity and directionality, and much of nature's diversity deviates from that linearity. As we've seen, there is branching all over the place. There is also the distinct possibility (in the abstract, but confirmed in practice) that evolution may, at least in specific cases, go from the complex to the simple: Asexual reproduction (in certain crustaceans, whiptailed lizards, and many other groups) seems to arise continually in many different lineages, which in itself is a form of simplification. And anyone who has seen the simple wormlike parasites in the gills of codfish is generally surprised when hearing, for the first time, that the parasites are copepods, whose free-living, non-parasitic relatives are very complex crustaceans. In other words, there is nothing in principle about the basic thesis that life has evolved that demands that evolution *must always* trace a path from the simple to the complex.

Yet one other set of predictions emanate, at least in principle, from the *scala naturae*. These are predictions about the distribution of anatomical characteristics, predictions that represent a revival of the terms *primitive* and *advanced,* which were more or less banned until recently. But here we run smack into the second of the grand, early patterns perceived in the spectrum of biological diversity: natural kinds. To see the positive effect the *scala naturae* has left on evolutionary biology, we must turn first to a defect of pattern perception. This flaw has haunted biologists from the days of Lamarck on, and it still bedevils some systematic biologists today.

In the preceding discussion of the *scala,* the word *group* or *groupings* occasionally appeared. I have said enough about "natural kinds" to indicate that both ancient and more recent forerunners of today's biological perspective delineated what they took to be real clusters of organisms, often with subordinate clusters. Linnaeus, a forerunner of Lamarck by a half century, has established with great success and acclaim his *Systema Naturae* in 1758. In it, clusters of species are associated in nested sets—in other words, in hierarchical fashion—with kingdoms holding classes which contain orders, then genera and then, finally, species.

Aristotle's penchant for dichotomies[6] (for example, blooded/blood-less), and its manifestations in the early days of systematic biology as invertebrate/vertebrates) has carried right down to the present time. For example, I work in a Department of Invertebrates, meaning a department assigned the awesome task of unraveling the history, and caring for vast collections, of all animals save the vertebrates. (And save insects, spiders, and scorpions as well; they are also invertebrates, but terrestrial arthropods are traditionally the purview of a separate department.) It is obvious that the word *vertebrates* means something, namely that the organisms are defined and recognized by the common possession of a vertebral column. Invertebrates, in sharp contrast, are all those animals *who lack a backbone.* Similarly, there is a dichotomy between the *coelomates* (animals with true body cavities) and the *acoelomates.* Within the (non) group acoelomates, the distinction is drawn between those animal phyla with true organs and those (like the coelenterates) that possess only tissues. Finally, the tissue-grade organisms contrast with the non-tissue-bearing animals such as sponges.

Take another important example: the dichotomy, often presented as the most fundamental in all of life, between the *prokaryotes* and the *eukaryotes*. We are eukaryotes, as are rosebushes, mushrooms, green algae, and amoebae: The cells of all these disparate organisms have well-defined nuclei housing the DNA, which is organized into chromosomes. Outside the nucleus lies a battery of *organelles*, some of which (the mitochondria of animals and heterotrophic microbes; the chloroplasts of plants and algae) are the cell's energy-transforming elements, each with its own single strand of DNA. Prokaryotes (that is, bacteria), in contrast, lack discrete nuclei, chromosomes, and organelles.[7]

Though we may indeed owe to Aristotelian dicta this curious penchant to divide the living world into two camps, the haves and the

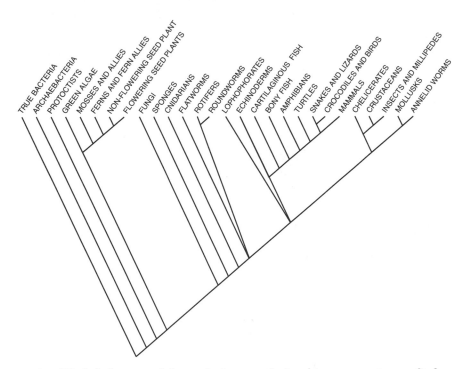

A simplified cladogram of the evolutionary relationships among (most of) the major divisions of extant life. The animal phyla are to the right of the fungi. Thus fungi and animals are considered each others' nearest relatives ("sister taxa") and as a whole are sister taxa of the group green algae + plants. These three multicellular eukaryote groups are, in turn, the sister taxa of the vast array of single-celled eukayrotes, represented here as the single line labeled "protoctists." The "prokaryotes" (true bacteria and archaebacteria) complete the phylogenetic array of life.

have-nots, I think a simpler explanation is in order: the *scala* simply begs to be sliced up this way. Lop off the top part of the animal sector of the *scala* and you have vertebrates — and everything else simply isn't a vertebrate. Lop off the eukaryotes, and you lump everything else into the "not-karyotes" [which is nearly the literal sense of the *pro* in prokaryotes, meaning "before (true) cells"]. As I have already mentioned, lumping all nonvertebrates together obscures the branching off of the protostomes and deuterostomes (each of which, for the record, are defined by *positive* attributes). It also obscures the branching off of a host of other, nonlinear structures within the animal branch of living diversity. And, lest this particular lopping seems top-heavy, there are a lot more bacteria than there are eukaryotic microbes, fungi, plant, and animal species combined. More to the point, some biologists claim that the two divisions of the bacteria — the *Eubacteria* (*eu* is Greek for "whole" or "true") and the *Archaebacteria* — is a more fundamental division of life than the have/have-not split between eu- and prokaryotes. Yet this division is wholly masked by lumping the two kinds of bacteria into the have-not prokaryote side of the fence.

A case can be made for the "naturalness" of a group like mammals. All mammals have hair and mammary glands. All but the spiny anteater (*Echidna*) and duckbill platypus have three middle-ear bones, and most of the living mammals have placentas, the exception being the marsupials (whose offspring complete their development in their mother's pouch), and the egg-laying monotremes (the aforementioned spiny anteater and platypus). When it was discovered that whales and dolphins likewise have hair (on the newborns, at least), three middle-ear bones, four-chambered hearts, and placental development, they too were recognized as mammals, not the fish they so closely, albeit superficially, resemble in external form.

For the most part, mammals (often referred to as "animals," even in highly literate publications) seem to go together, as if all were variant versions of the same sort of creature. The pattern of association by similarity is even clearer when we look at smaller subsets of mammals, such as cats (the family Felidae in modern classifications). True cats are by and large so stereotyped in behavior and basic body shape that the major difference between species often seems mostly a matter of overall body size — at least to the casual observer. All cats seem naturally to go together, to form a "natural" group.

As Mayr has pointed out (see note 6 to this chapter), Aristotle's classification of animals reflects this sort of association through similarity far more than it does his theoretical penchant for the sort of dichotomizing into haves and have-nots. But it was of course Linnaeus who really got the ball rolling on the hierarchical classification of the living world. Though Aristotle recognized subdivisions of groups, it was Linnaeus who first propounded a formal system that recognized that the living world seems to consist of nested sets of organisms.

Though Linnaeus's main forte was botany, it is perhaps simpler to characterize his system by first considering how Linnaeus himself, as a member of the species *Homo sapiens* (a species which Linnaeus actually named), is classified. Humans, of course, are animals. Within the kingdom Animalia, we are vertebrates—which in formal terms comprise the subphylum Vertebrata of the phylum Chordata. We share with other animals hair and mammary glands, making us mammals (class Mammalia). Along with tarsiers, lemurs, lorises, monkeys, and apes, we are primates (order Primates). With some extinct genera (for example, *Australopithecus*), we are members or the family Hominidae, genus *Homo*, species *Homo sapiens*.

The Linnaean categories shown below are precisely that: names of the ranks arranged from highest to lowest. *Instances*, meaning particular sets of organisms, allocated to one or another rank (such as the species *Homo sapiens*[8] or the phylum Mollusca) are what modern systematists call *taxa*—what the early naturalists thought of as "natural kinds," and what post-Darwinian systematists have come to think of as groups of species that are evolutionarily united (by commonality of evolutionary descent). Thus the ranking categories themselves form a *scalar* hierarchy (lowest to highest), but the actual groupings of species, the actual taxa, are nested like a set of Chinese boxes, forming a *subordinated* hierarchy.

The Linnaean classification of the species *Homo sapiens*. Categories are ranked, while taxa (specific instances) are nested within one another. Thus, *Homo sapiens* is a species within the genus *Homo*, and *Homo* a genus of the family Hominidae, and so forth.

Category	Taxon
Kingdom	Animalia
Phylum	Chordata
Subphylum	Vertebrata
Class	Mammalia
Order	Primates
Family	Hominidae
Genus	*Homo*
Species	*Homo sapiens*

The term *Linnaean hierarchy* thus refers to two separable hierarchical components.

Biologists from Linnaeus on down have sought to describe and, in effect, "define" taxa using the attributes of organisms themselves. For the most part, they have gravitated to as few as one, or as many as several dozens, of characteristic features that seem to pull a group together. One early botanist, Michel Adanson (1727–1806), actually did suggest that groups be recognized by taking absolutely all features into account. This would seem to maximize the "association by (overall) similarity," which is the earliest and perhaps most intuitive approach to recognizing "natural kinds." Actually, Adanson was most concerned that all potential features be examined in the search for natural groups—in lieu, that is, of assuming that only certain parts of an organism (like the flowers or fruits of an angiosperm plant) will reveal the natural order.

But even the earliest botanists and zoologists knew that, somehow, some characters were more informative than others in revealing the orderly relations between organic beings—the nested patterns of "natural kinds." For example, mammals being four-legged vertebrates, it would be useless to cite "four-leggedness" or "presence of vertebral column" to define mammals, as the former also embraces amphibians, reptiles, and birds (wings are modified forelimbs), and the latter encompasses cartilaginous and bony fishes as well. It is best to cite as definitional only those features that are limited in their distribution solely to the "natural kind" being defined: hair and mammary glands for all mammals, or three middle-ear bones for marsupials and placental mammals, or placentas for the eutherians (placental mammals).

Gareth Nelson,[9] my colleague for many years at the American Museum, had a rich insight into the techniques of cladistic analysis (briefly encountered in Chapter 1). He posited that they are nothing more or less than the codification of sound practice in systematic biology that goes back at least to the days of Adanson and Linnaeus. An example from my own area of invertebrates, and one of my favorite scientific papers, was written by British Museum zoologist E. Ray Lankester in 1881. Lankester was writing after the Darwinian revolution was completed but over a half century before Hennig articulated the principles of phylogenetic systematics. Entitled, simply, "*Limulus* an Arachnid," Lankester's paper tersely lists a number of features of the modern-day horseshoe "crab" that, he felt certain, showed that horseshoe crabs are not true crustaceans but rather are allied with spiders and scorpions. For example, instead of antennae, horseshoe crabs have a pair of short legs bearing pincers at the end just in front of the mouth. These appendages are very similar to the ones in the same place in scorpions and spiders. In modern terms, Lankester was saying that the features unique to scorpions and spiders are also found in horseshoe crabs; also, those that define crustaceans, including larval characters and the presence of two pairs of antennae, are utterly absent in horseshoe crabs. Lankester was simply correcting a mistake by pointing to features previously overlooked. In principle, both Linnaeus and Adanson, had they been apprised of the facts of the matter, would have done the same thing: They would have realized that horseshoe crabs are similar to true crabs mostly in their common habitat: Both live in the sea and extract oxygen from seawater by gills, a feature which is not especially similar in the two groups.

Certain groups, like mammals or flowering plants, are obvious to the eye, easy to recognize, and simple to characterize in terms of their unique, intrinsic features. The task of cladistics is to devise ways of dealing with the less obvious groups. And here again we intersect the *scala naturae*. As we have already seen, it is a relatively simple matter to lop off the tops of individual subsets of the *scala*. (We would now, in post-Darwinian times, refer to subsets as separate lineages or *clades*.) Thus, the array would be rent into the positively defined haves versus a motley contingent of have-nots, the latter "defined" only by the *absence* of the features that define the coordinate "have" group. Groups that are difficult to name lack obvious defining features. These groups usually turn out to be the dregs of a lineage, after all the evolutionary highly modified, and thus uniquely stamped, taxa have been culled. All those have-not invertebrate "groups" don't really exist as such. But the problem remains: How do we recognize pattern, nested sets, within such lineages as the lower half of the grand animal clade, say all those taxa (phyla, in this case) below the "body cavity" level?

Cladistics, in keeping with the original search for natural kinds, but imbued with an evolutionary perspective, boils down to a search for the actual distribution of the features, or "characters," of organisms. It turns out that such characters are complexly, hierarchically nested just as are the taxa they define. Hennig's expression for this pattern is hardly vernacular: the *heterobathmy of synapomorphy*. (Small wonder he needed Lars Brundin, and later Gareth Nelson and others, to make his ideas clear!) Yet Hennig's expression actually hits the nail on the head: *Heterobathmy* means "different depths"; *synapomorphy* means those features unique to a group. Hair defines mammals, is present in one form or another in all mammals, and is found in no other group. Scales, however, are primitive versions of hair and of feathers. So "scales" in a sense includes hair and feathers, and the term is useful to define a group containing reptiles, birds, and mammals. That group is also defined and recognized, conveniently and obviously enough, by the presence in all three subsets of the amniote egg. Scales are at a deeper, more inclusive "depth" of distribution than either hair or feathers. Hence the heterobathmy of synapomorphy. Scales are the early expression of epidermal features, hair and feathers, that we now understand arose later, as modifications of those scales. Hence it is perfectly objective, proper, and appropriate to label scales, such as those seen in a rattlesnake skin, as "primitive" vis-à-vis the advanced (or "derived") condition of a rat's hair or an ostrich's feathers.

Pre-Darwinian biologists saw that characters such as hair, feathers, and scales were every bit as "nested" as were the natural kinds they delimited. The concepts of homology and analogy predate the Darwinian revolution, as earlier comparative anatomists recognized that some characters resemble one another because they are in some sense the same, (homologous) whereas others are only superficially similar, and not really the same (analogous). Thus the vertebrate forelimb (for example, the wings of birds and bats, human arms, the front legs of frogs and lizards, and the pectoral fin of bony fishes) all bore some underlying common structural pattern and are thus *homologous*. In contrast, the wings of birds and insects are merely *analogous*, both enabling flight. And, perhaps confusingly, the wings of bats and birds, homologous as vertebrate forelimbs, nonetheless differ in basic plan as wings, so therefore *as wings* they are analogous.

The point of all this, as we shall see in greater detail when considering the Darwinian revolution, is that nested sets of homologous features turned out to be readily understood as Hennig's "heterobathy of synapomorphy"—and as a natural and necessary consequence of the basic process of biological genealogy: evolution.

The relation between nested sets of organisms—natural kinds in pre-Darwinian terms, simply *taxa* in evolutionary terms—and the *scala naturae* leaps out of a "cladogram" of the relationships between the various thirty-six or so invertebrate phyla. The animal section of Lamarck's *scala* (and Aristotle's for that matter) survives almost intact. But within it we see the distribution of characters, and we see that characters that define a large sector of the diagram (the synapopmorphies for that sector) immediately become primitive as one goes on up the scale, within that group. Thus the feature survives in its original form, and becomes modified (sometimes almost beyond recognition) in other groups. Hair helps define mammals, feathers define birds, *but scales do not define reptiles* because they are primitive holdovers. As we have seen, scales help define the group Amniota, consisting not only of reptiles but also of birds and mammals.

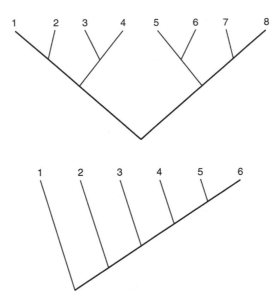

Two hypothetical cladograms. On the top, the evolutionary relationships
among eight species, forming a subordinated pattern of symmetrically nested
sets, is shown. On the bottom, a pectinate pattern of relationships among six
species — reminiscent of the *scala naturae* pattern emphasized by Lamarck. As
seen in the cladogram on page 66, there is a preponderance of pectinate pattern in the history of life.

In the twenty-five-odd years since the methodology of cladistics became the approach of choice throughout the world of systematics, one
thing has become striking. Despite the presence of a number of symmetrically nested cladograms, by far the most common result takes the form
of a *pectinate* (that is, "comblike") structure, where branch after branch
comes off a single ascending line, each notch defined by what is now recognized as an additional "evolutionary novelty" as shown in the figure
above.

Thus, in the great and profusely diverse array of life on earth, there
are both *scala*, a range of primitive-derived features obvious to the earliest observers, and nested, inclusive sets of organisms. These are two sorts
of patterns, both fraught with evolutionary implications. Lamarck argued
that the *scala* implies that evolution *has* occurred, that some process of
ancestry and descent interlinks all life. His arguments didn't prove utterly convincing, even though he also included a speculative notion of
how evolution might have occurred. It would seem, then, that it is the
less-than-precise correlation between prediction and observation from

early considerations of the scala that in the end made Lamarck's work not entirely convincing to his colleagues. Yet, today, in the heyday of cladistics, the *scala* emerges in one pectinate cladogram after another, silent testimony to the connection between evolution and this grand natural pattern. This, even though the pattern itself was not able to force itself sufficiently on the minds of the early biologists (save Lamarck!) to convince them that life had in fact evolved.

Yet that is not the whole story. Lamarck, simply by arriving earlier than Darwin, faced a world more entrenched in the notion that whatever order one might discern in the living world is a reflection of God's handiwork rather than natural process. Linnaeus steadfastly refused to embrace anything like an evolutionary stance, though late in his career he seemed to accept a form of species changing from one to another, albeit within the context of the immutable genus. His overwhelming message was embodied in his famous phrase *"Species tot sunt quot ab initio faciebat Infinitum Ens"* ("There are as many species as originally created by the Infinite Being"). To Linnaeus, there is indeed pattern in biological nature, but it is pattern fashioned strictly by God. Charles Darwin did not agree.

Darwin, Patterns, and Evolution

Charles Robert Darwin (1809–1882) wrote a number of books and monographs, along with many shorter papers, over his long career. All are admirable and valuable, but none of the others remotely approaches the significance of the "abstract" of his ideas, published in 1859 as the first edition of *On the Origin of Species by Means of Natural Selection, or the Preservation of Favoured Races in the Struggle for Life*. It was Darwin's task in this, his most famous volume, simply to convince the thinking, literate world[10] of the *fact* of evolution—that, as he put it in his last chapter (1859, p. 484), "probably all the organic beings which have ever lived on this earth have descended from some one primordial form."

In this task he was wholly successful within the framework of western science. With the appearance of the *Origin*, it became virtually impossible for professional biologists to continue to attribute the diversity of life to the direct actions of a supernatural Creator. Rather, regardless of what one's personal views of the nature and fundamental actions of God might be, it became de rigueur in biology and paleontology to invoke what Darwin himself called "secondary causes," which he proclaimed "accords better with what we know of the laws impressed on matter by the Creator" (p. 488).

Though Darwin referred to his goodly sized volume as an "abstract," he also termed it "one long argument,"[11] which it truly is. Darwin was at great pains to reveal what he simply termed the *facts* of natural history — a word which can safely be considered a virtual synonym of *patterns* as used throughout this narrative. True, the *Origin* is crammed with specific examples drawn not only from Darwin's own observations of the natural world but also from the work of preceding and especially contemporary zoologists, botanists, anatomists, paleontologists, and geologists. But each example is a specific instance of what we are always to take as a general set of such instances. All affirm one or another basic regularity of the natural world which Darwin musters to convince the reader of the truth of his fundamental proposition: that life has evolved. Darwin's *Origin* is indeed an "abstract" of a much-longer planned work, but there is no doubt that the more voluminous and never completed volume differs from the *Origin* solely in the number of specific examples intended. The *Origin*, with its fewer but still impressive examples, nonetheless exposes all the grand patterns with which Darwin hoped to convince his readers that life has evolved.

Darwin's modus operandi is simplicity itself and wholly modern in its logical basis of scientific argumentation.[12] For Darwin in effect is always asking us to imagine what we would expect to find in the natural world if evolution were true. His primal pattern, as we shall soon see, is the Linnaean hierarchy of "natural" taxa. These taxa, he argues in three separate sections of the *Origin*, *must* be there if life evolved. This is actually a subtle form of the famous *hypothetico-deductive method*, wherein scientists are supposed to predict phenomena they might observe, whether in experimental or natural conditions, *if* their conjecture (their "hypothesis") about some specific aspect of the nature of physical entities or interactions is correct. Generally it is claimed[13] that under the hypothetico-deductive method, the ultimate truth of a proposition can logically never be proven. Instead, we are able only to reject a proposition if we fail to observe its predicted consequences.

Recall the discussion in Chapter 1, where I argue that there is in fact no logical distinction to be made between functional (especially experimental) and historical science. I reiterate here that predictions can apply to patterns we may expect to find in the living world every bit as much they apply in more stereotypically "scientific" experimental situations. Likewise, they may apply to the geological record of both the physical history of the earth as well as of life itself. Such, in any case, was Darwin's proposition. He did not just reinterpret the "facts" of natural history. Rather, he proclaimed that those variegated patterns were necessary consequences, confirmed predictions, all of which must be true if the central proposition that life had evolved were itself true.

Moreover, Darwin was forever raising counterarguments to his own thesis. He does so throughout all his chapters, and not just in the original Chapter 6 (1859) of the *Origin*, "Difficulties on Theory." He was aware that one pattern confirming his thesis was not enough. Other explanations were already available, especially those involving supernatural acts of creation. It was, rather, the preponderance of evidence, the long string of patterns that *must* be present if evolution had occurred, that Darwin relied on to establish the general plausibility of his thesis. One pattern confirming the hypothesis that life had evolved might be mere coincidence, but, say, 7 or 10 such confirming patterns (depending how one counts up Darwin's major and minor patterns), that's another matter entirely. The probability that all such disparate predictions would hold and yet the proposition of evolution *not* be true becomes so small that we can accept with confidence (or as "fact"; see note 10) the hypothesis that life has evolved.

If there is a single coherent theme interlinking the otherwise disparate patterns Darwin invokes to support his argument, it is *connectedness* or *continuity*. "Descent with modification," the transformation of an ancestral species into a descendant, *means* connectedness. Darwin was out to refute the entrenched, ambient attitude that was never more succinctly nor trenchantly put than by the much-respected William Whewell, who wrote (in his *History of the Inductive Sciences*, 1837, p. 626) that "species have a real existence in nature, and a transition from one to another does not exist." It is clear that Darwin felt that he had to deny the existence of species, that is, the validity of the first of Whewell's cited propositions (see "Pattern Denied," on page 84), in order to successfully deny the truth of Whewell's second proposition. But it was primarily Darwin's citation of pattern after pattern which demonstrated a connectedness, which established that "a transition" between species does indeed "exist."

What, then, were these patterns that Darwin so cleverly and successfully invoked as necessary consequences of the genealogical process of "descent with modification"? They are the very list that many of us had to memorize in school (where evolution was taught!). Such is Darwin's "evidence" for evolution: geographical and geological distributional patterns; intergradational patterns, of species and subspecies and of hybrids; embryological patterns; and the closely related morphological patterns which Darwin folded neatly into the grandest pattern of all: the Linnaean hierarchy. The *scala naturae* is mentioned only in passing, the more remarkable given Darwin's penchant for seeing progress and improvement in evolutionary history. I cover here only what have emerged as the most essential of Darwin's patterns. Omitted, for example, are his separate-chapter treatments of hybridization and instinct. The essence of his arguments on continuity and connectedness lies especially in his treatment of geographical and geological distributions (4 of his 14 chapters) and above all else, his treatment of the Linnaean hierarchy.

Pattern Applied

The first four chapters of the *Origin* are themselves a linear argument leading up to the enunciation of the "principle" of natural selection (see page 89). But they also serve as a proclamation of the continuity between variation within a species, the existence of geographic variants (variously termed *races* or *subspecies*) within a species, and the presence of closely related species living in adjacent regions. For example, at one stage Darwin invokes the red grouse, which he says some observers think is the same species in both Norway and England; others, he states, believe them to be specifically distinct, the British form being a locally endemic separate species from the mainland European red grouse. If you can't tell whether two geographic "forms" are just geographic variants ("races") or separate species, you have a prima facie case for concluding that there is a sliding scale between geographic differentiation *within* a species and the eventual production of separate species from a common ancestor. And that, as Ernst Mayr was to note many years later,[14] is precisely what one would expect, indeed *predict*, from the general notion of evolution.

Darwin on the Geographic Distributions of Plants and Animals. Darwin first addresses specifically geographical distributional patterns in Chapter 2 of the *Origin*, in conjunction with his linear argument developing the very concept of natural selection (see page 89). Darwin picks up the theme again later in two additional chapters (11 and 12 of the first edition). Evolution, to Darwin, implies point origins of species. For example, in his summary of the geographic patterns chapters, Darwin writes "with respect to the distinct species of the same genus, which *on my theory* [my italics] must have spread from one parent source. . ." (*Origin*, 1st ed., p. 407). (Talk about observations predicted from theory!) Hence, Darwin goes to great lengths to establish two basic geographic patterns.

First, Darwin noted, species in far-flung regions of the globe tend to be related to other species living in the same region rather than to species living in the same general habitats (alpine meadows, deserts, and so on) in far-removed places. That's why there are grand patterns of geographic distribution, with entire families, even orders, restricted to certain regions of the globe. Long periods of evolutionary diversification in isolation should lead to the development of areas of *endemism* (the restriction of a number of species to a relatively narrow region), and to major regional biotas. The Cape flora (*fynbos*) of Southern Africa, which was cited, if only in passing, by Darwin, remains an excellent example of a small, relatively localized area of endemism. It consists of just the rim of the narrow southernmost margin of the African continent. But it contains such a profusion of unique plants that it ranks as one of the six major floral regions of the world!

Islands constituted the second main geographic pattern Darwin cited. He proclaimed that their lessons transcend the narrow interpretation of oceanic outposts, applying as well to mountaintops and other spottily distributed environmental systems. Island floras and faunas often appear to be peculiar subsets of the variety of life on the closest continental regions, and often they have progressively fewer different groups than otherwise might be expected the farther a particular island lies from the mainland.

Tigers, for example, go as far as Bali and stop; the top predator on Sulawesi (the former Borneo) is the rock python, not tigers or any other mammalian species. But islands offer far more than mere absences: Species are often endemic to particular islands. And island chains, such as the celebrated Galapagos, often have closely related but clearly distinct species on the different islands within the chain. Just as one might expect, "under my theory," Darwin claimed that isolation promotes divergence, and as species spread from island to island within a chain, one would predict the process to continue—for species to diverge and ultimately become separate on the different islands. Most famous are the Galapagos finches.[15] But the tortoises too show a very clear pattern of differentiation from island to island, although they are generally considered to be subspecies of a single species.

My own favorite example—also known to Darwin, but not in modern detail—are the Galapagos mockingbirds, of which there are four species. The "Galapagos mockingbird" occupies all the main islands *except* the three islands to the extreme southeast of the chain: Floreana (known as "Charles" to Darwin, hence, the "Charles mockingbird" of Floreana), Hood (the "Hood mockingbird") and San Cristobal (also known as Chatham; hence the "Chatham mockingbird"). Darwin would have been delighted to know that absolute dating reveals these three islands to be the oldest of the Galapagos. He would have predicted that the most highly differentiated species—as in the three separate mockingbird species restricted to single islands—would occur on the oldest islands of the chain. As it was, all he knew was that the species on

Chatham and Charles are distinct "mocking-thrushes," as he called them.

Indeed, since Darwin's day the science of *biogeography* has changed. This analysis of the ecological and evolutionary components of the patterns of distribution of plants, animals, and other elements of the biota over the earth's face is no longer just an empirical mapping exercise, as it was to most of its adherents in the 1800s. Plate tectonics, for example, has greatly modified our understanding of why species occur where they do: Ostriches and their kin (rheas of South America, emus of Australia) are restricted to the continents of the southern hemisphere, a pattern of distribution that suggests that those continents might formerly have been closer together. Such a pattern is very much in keeping with Darwin's realization that closely related species tend to occur in localized single regions. Indeed, and as we shall see in greater detail in the following chapter, probably because scientists denied that those continents were actually connected into the supercontinent of Gondwana, ornithologists tended to regard the ostrichlike birds as independently evolved flightless species, each adapted to open, semidesert environments. Only after plate tectonics established the undoubted past reality of Gondwana were ornithologists able to muster the evidence to convince themselves that the ostrichlike birds of the southern hemisphere are in fact a natural, evolutionarily coherent single group! Thus, even in the face of plate tectonics, Darwin's basic take on the meaning of patterns in the distribution of life over the globe remains solid.[16]

Darwin on the Geological Distribution of Fossils: Positive Patterns

It is striking how seldom in the *Origin* Darwin refers to Cuvier, Lamarck, or even Linnaeus. No mention is made explicitly of the great "chain of being" that was the cornerstone pattern cited by Lamarck in support of the very notion of evolution. It is near the end of the second of two chapters devoted to the geological record of life that Darwin comes closest to discussing the *scala*, and he does so with some initial trepidation:

> There has been much discussion whether recent forms are more highly developed than ancient. I will not here enter on this subject, for naturalists have not as yet defined to each other's satisfaction what is meant by high and low forms. But in one particular sense the more recent forms must, on my new theory, be higher than the more ancient; for each new species is formed by having had some advantage in the struggle for life

over other and preceding forms. If under a nearly similar climate, the eocene inhabitants of one quarter of the world were put into competition with the existing [that is, *modern*-NE] inhabitants of the same or some other quarter, the eocene fauna or flora would certainly be beaten and exterminated. . . . (Darwin, *Origin*, 1st ed., pp. 336–337)

Thus Darwin eludes the issue of anatomical complexity, and even the notion of progress, defining *higher* only in the special sense of later faunas and floras being able to outcompete older ones. Though this indeed is the traditional interpretation of the extinction of the marsupial and eutherian mammals of South America (that is, that they were faced with and exterminated by invading mammals from the north once the Isthmus of Panama was uplifted and consolidated into a terrestrial corridor), more recent interpretations have abandoned this general line of thinking. Indeed, there is little reason to suppose that a modern marine invertebrate fauna would in fact be able to dislodge its Eocene equivalent.

Beyond Darwin's untestable and not altogether plausible conjecture of the superiority of progressively later faunas and floras, his use of the supposed progression of floras and faunas melds in with his more general statements of patterns of the fossil record. Relying, for example, on Charles Lyell's analysis of Tertiary marine invertebrate faunas (mainly molluscan), Darwin argues that they become distinctly and progressively more like the modern faunas as one samples Tertiary marine formations from the oldest to the youngest. In this, Lyell and Darwin were entirely correct.

Another element to this pattern, Darwin notes, is that faunas above and below one another (*subjacent*) will bear the closest resemblances, while faunas temporally more remote from one another will share fewer resemblances. By "more similar," Darwin for the most part simply means species in common, or more generally species very similar to one another, as, for example, members of the same genus or family. To put it in late twentieth-century terms, faunas that are more remotely spaced in time will have comparable "ecological roles" played by members of different families, or even orders. Here again, Darwin was of course right to suggest that such patterns were "consistent" with "my new theory."

Thus the core of Darwin's geological patterns cited in confirmation of his notion of descent with modification is once again continuity and connectivity. But we have yet to confront perhaps his most powerful argument on geological succession: the one (Chapter 10, p. 301 ff.) that is a mere variant version of his treatment of the Linnaean hierarchy, presented first in Chapter 4 on "Natural Selection" (p. 116 ff.). The subject reappears yet a third time, in Chapter 13: "Mutual Affinities of Organic Beings: Morphology; Embryology; Rudimentary Organs" (p. 432 ff). For in these disparately titled sections we finally confront nothing less than Darwin's main pattern, his clinching argument that life *must* have evolved, all centering around *"Darwin's diagram."*

Darwin's Diagram: Evolution and the Linnaean Hierarchy

The *Origin* has but one illustration, yet seldom has a single diagram served such powerful purposes. Shown here on pages 82 and 83 it was published as an interleaf between pages 116 and 117 of Chapter 4 of the *Origin,* where it was first discussed. The diagram is a welter of straight horizontal solid lines and off-vertical interconnected solid or dashed lines. The symbols, such as the capital letters A to L near the bottom, have various meanings, depending on which of the three separate discussions is consulted. For example, they denote *"species"* in Chapter 4, while they represent Silurian genera in Chapter 13. Similarly, in Chapter 4 the horizontal lines stand for the passage of 1000 generations, whereas they stand for the passage of geological time in the subsequent two discussions.

However differently motivated, and however the passages differ in specific details, the import of the diagram in all three instances is identical: Darwin is arguing that, if his theory of descent with modification is true, and given the possibility of divergence (the splitting of lineages, there being obviously far more than a single species now living), *then it must also be true that a nested pattern of species clustered into ever greater assemblages will automatically result.* In other words, and using modern terminology, so long as the system admits the splitting of lineages, evolution *must* produce a nested, thus hierarchically arrayed, complex of taxa.

We have already seen the reason why this is so in the discussion in Chapter 1 on reconstructing genealogies in fields as disparate as biological systematics, historical linguistics, and stemmatics (the history of handcopied manuscripts). And, as we have seen in this chapter, pre-Darwinian biologists had recognized patterns of greater and lesser similarity among organisms based on their attributes. Such anatomical, as well as behavioral and physiological, characters are now known collectively as the *phenotype.* Hair, feathers, and scales are *homologues,* with scales being the more general, primitive condition, hair and feathers the more advanced or "derived." The core question of evolution has always been the diversity of life, meaning, at base, *phenotypic* diversity (now understood explicitly to have a genetic base: *genotypic* diversity).

Thus, Darwin claimed in all three discussions that as time passes, attributes change. Natural selection acts on heritable variants, which is especially discussed in detail in the first of Darwin's three cited passages. As lineages split, and further change accumulates, what happens in one lineage has no effect in collateral lineages. Modifications are handed down in the same or further modified form only to direct descendants. All genealogical systems behave this way, so long as there is splitting of lineages. This behavior results in a branching array of relatively more closely related lineages. In short, a nested set of clusters of species is produced that is defined and recognized by a coordinate set of nested homologies—Hennig's *synapomorphies*.

This, really, is evolution's grand prediction. One would expect to find general sorts of intergradations among species geographically, and among species and entire floras and faunas, as one moves laterally or up and down the geological column. But these patterns are difficult to specify with precision. Likewise, the *scala naturae*: Starting with the simplest forms (prokaryotic bacteria), one might suppose that additional complexity would be the only direction in which the evolutionary system could go. One might even imagine a ratcheting mechanism, where life-forms of greater complexity occasionally appear and survive, not, as Darwin supposes, because the "higher forms" outcompete and drive to extinction the "lower forms," but for precisely the opposite reason. The higher forms do not supplant or otherwise compete with their earlier, simpler forebears. Hence they improve their chances for survival and allow a sort of stepwise cumulative "ratcheting up" of complexity as life evolves. Yet this is also vague and certainly not a *necessary* prediction of the mere supposition of "descent with modification". There is nothing inherent in the "rules" of evolution that says categorically that life *must* have progressed beyond the simplest bacterial level. This Darwin himself saw very clearly, hence his reluctance to cite *scala*-like patterns in support of his "new theory."

That's the power of Darwin's insight: that "natural kinds," nested taxa, are a necessary consequence of "descent with modification." That the pattern holds was made clear by Linnaeus a century earlier. The empirical evidence preceded the evolutionary interpretation. So, too, did empirical evidence for faunal and floral turnover, which served as the basis for the development of the geological timescale, *also* predate interpretations of evolution and extinction. The Linnaean hierarchy is the *only* utterly unambiguous pattern that the simple idea of descent with modification unequivocally predicts.

There are some interesting nuances in each of Darwin's three discussions. One that is noteworthy here is the "on first principles" imaginary course of evolution through natural selection, which is applied in

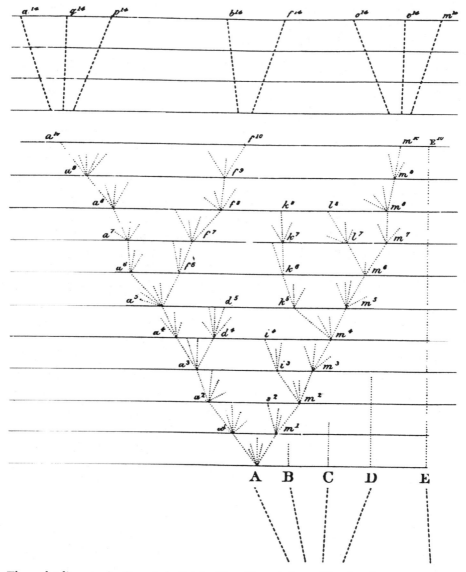

The only diagram in Darwin's *Origin*. Note Darwin's predilection for symmetrically nested sets over pectinate patterns.

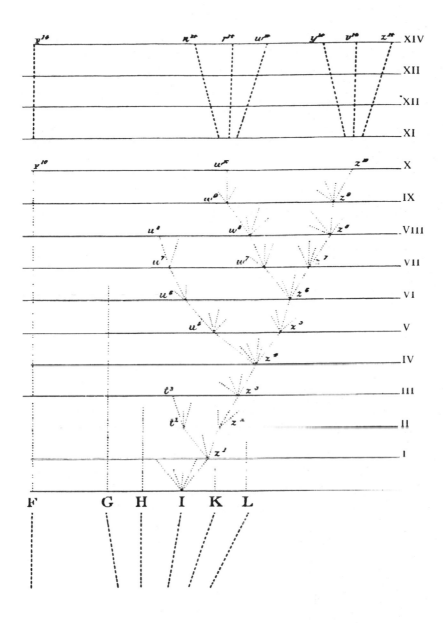

discussing the diagram when it is first encountered in Chapter 4. Here Darwin relies heavily on his "principle of divergence," wherein, he says, natural selection tends to change the phenotypic properties of organisms away from the parental "type." Here, the emphasis is on how selection can transform those properties. In contrast, little is said of selection in the two later invocations of the diagram. Indeed, it is nested patterns of resemblance of anatomical characters (including embryology, where embryos of different species tend to resemble one another progressively more closely the earlier on in development one looks) that form the focal point of his third analysis.[17] However, it is perhaps the clearest and most articulate linkage of evolution with its main, necessary consequence: the production of nested sets of taxa, which is appropriately spelled out in a section on classification.

Does the existence of the Linnaean hierarchy *prove* that evolution is *true?* Technically, the answer is: Of course not. It is possible that other agents and processes could have created natural sets of taxa. But in another sense, the answer is: Sure it does. Because, other than spontaneous generation, there have never been nor are there now any other natural, material, and hence "scientific" principles brought forth that might account for the nested pattern of resemblance that links up the astonishingly diverse array of life on earth.[18]

Most of Darwin's contemporaries became convinced virtually overnight that life must have evolved. The world might have been primed and waiting to accept something they already knew or suspected to be true. There were, of course, persistent skeptics (mostly the older generation, as Darwin foresaw). But, by and large, Darwin succeeded in transforming the scientific establishment. He made it suddenly unfashionable, even impossible, in scientific circles, *not* to accept the basic proposition that all the earth's species are descended from a common ancestor in the dim reaches of geological time. How much the notion of natural selection played a role in this monumental mind change is still debated. (The subject is addressed in a bit more detail below.) But I am convinced that Darwin succeeded where so many before him had failed because he managed to pick the one compelling, unshakable pattern that simply *must* be obtained if his ideas were true. It a sense, then, it was a "done deal" from the start. Everyone knew the pattern existed long before Darwin took up his expostulatory cudgels.

Pattern Denied

Reading the *Origin,* I am persistently struck with Darwin's almost wholly, purely genealogical viewpoint. As a modern evolutionary biologist, I can only admire recognition of genealogy as the very essence of evolutionary thinking. It led him to see many patterns of continuity and connectedness, including especially the pattern embodied in the Linnaean hierarchy. Yet Darwin's strict allegiance to purely genealogical pattern and interpretation led him to overlook other very vital aspects of biotic pattern, ones that have been left to succeeding generations to address. In the nearly century and a half since publication of the *Origin*, patterns have emerged that are equally real, patterns that Darwin saw as either against or at least irrelevant to his theory.

Most notably, Darwin continually prefers to view competitive interactions between species as the cause of certain phenomena (especially extinction), as opposed to nonbiological, physical causes. The one exception to this otherwise general, and significant, observation comes in his development of the concept of natural selection (see page 89).

Along with, and perhaps related to, his preference for biotic versus physical explanation is Darwin's reluctance to embrace cross-genealogical phenomena. Given his insistence on solid first principles of genealogy one would *not* predict either the simultaneous appearance or disappearance of species. Such, after all, would only confuse the issue of genealogical transformation: descent with modification. This denial of cross-genealogical patterns of biotic history, at least as relevant to his argument that life has in fact evolved, is the first of two general categories of patterns denied. These patterns have re-emerged to preoccupy evolutionary biologists in the years since the *Origin* first appeared.

As to Darwin's devotion to pure genealogy, consider the critical passage at the outset of his second chapter. Here, in what amounts to a thumbnail summary of his theory, Darwin expounds on what the geological succession of life has to say about his theory.

> These several facts accord well with my theory. I believe in no fixed law of development, causing all the inhabitants of a country to change abruptly, or simultaneously, or to an equal degree. The process of modification must be extremely slow. The variability of each species is quite independent of that of all others. Whether such variability be taken advantage of by Natural Selection, and whether the variations be accumulated to a greater or lesser amount, thus causing a greater or lesser amount of modification in the varying species, depends on many complex contingencies. . . . Hence it is by no means surprising that one species should retain the same identical form much longer than others; or, if changing, that it should change less.[20] (Darwin, *Origin*, 1st ed., p. 314)

Darwin has the eyes of a genealogical anatomist, dissecting faunas and floras to their basal components: species. He would not insist that species do *not* affect one another (as for example with his theories on competition and extinction; see below). Nonetheless, he does see the evolutionary histories of species and species lineages as distinct and, in a profound sense, independent from one another.

Additional examples abound of Darwin's devotion to genealogy. In the brief synopsis beginning Chapter 10 of the *Origin* (the second geological chapter), Darwin states: "Groups of species follow the same general rules in their appearance and disappearance as do single species," perhaps suggesting a discussion of the simultaneous appearance and disappearance of *unrelated* species. Such were the patterns that led Cuvier to his notions of catastrophic extinctions and the subsequent emergence of new faunas, and also to the empirical construction of the geological timescale. But no: Turning to that discussion (p. 316), Darwin refers to "groups of species, that is, genera and families . . ."—meaning genealogical groups of species. Darwin was out to demonstrate the truth of his general proposition of descent with modification. To that extent, all patently nongenealogical pattern was either misleading or false. And in some instances it was to be explicitly denied.

Similarly, in "On the forms of life changing almost simultaneously around the world," Darwin renounces pure parallelism in favor of the spread of "dominant" species around the world, who are then likely to produce more dominant descendant species of like kind. Whatever the title of the section might suggest, Darwin the strict genealogist always adopted pattern explanation that emphasized continuity of descent rather than the wholly independent production of confusingly similar, yet unrelated, forms in disparate sectors of the globe.

In other words, everywhere that the modern reader might anticipate a discussion of cross-genealogical, or *ecological*,[21] patterns, Darwin makes it clear that when he is speaking of multiple species, they are all related. Nowhere do we see addressed the sorts of cross-genealogical extinctions documented by Cuvier, the phenomena that served as the empirical basis for constructing the geological timescale. Darwin expresses his "patterns denied" in two general ways: explicitly, especially with Chapters 6, "Difficulties on Theory," and 9, "On the Imperfection of the Geological Record"; and implicitly, by ignoring them completely. The absence of any discussion of Cuvierian patterns of origin and extinction of multiple components of entire faunas and floras is striking. Once again, after seeing a headline at the beginning of Chapter 9 (p. 279), "On the sudden appearance of groups of species," we think that such a discussion might be found. Instead when we refer to the heading where the material is discussed in the chapter, we find two words added, the latter significant and by now quite predictable: "On the sudden appearance of *whole* groups of *Allied* Species [italics mine]." Such a pattern, where there is no evidence of time elapsing between appearances of related species, is indeed problematical to Darwin's notion. And he attempts to alleviate it by claiming that the fossil record is inherently poor, and our experience (circa 1859) insufficient for the *real* genealogical signal to emerge.

Thus I find the absence of cross-genealogical pattern not so much reflecting a difficulty for Darwin, who was striving so mightily to answer all his critics' rejoinders in advance, than a nonissue. These patterns could safely be ignored by Darwin because while on the one hand they did not support his argument for a genealogical history of life, on the other they did not raise fatal objections to it. The patterns that did seem to pose difficulties were those that appeared to argue against a regular pattern of ancestry and descent, such as the apparent simultaneous appearance of *related* species. Yet we will find that the elusive link to the physical world, the world of matter-energy transfer, is largely through ecosystems. And such systems are typically composed of populations drawn from many different and decidedly not closely related species. Darwin's sole memorable mention of cross-genealogical systems comes in his final paragraph, where he finds it "interesting to contemplate an entangled bank." It is easy to criticize even an intellectual giant like Charles Darwin for omissions. However, these were less sins than logical omissions in his desire to simply establish the plausibility of genealogical descent with modification. But, in so omitting cross-genealogical pattern, Darwin did leave it to those of us who have followed him to forge stronger, more realistic links with the world of matter-energy transfer processes than he himself managed to do.

The second general category of "patterns denied" are any data, real or artifactual, that speak against continuity and connectedness. These Darwin tended to pinpoint and attack vigorously, claiming in the main that such patterns are more imagined than real—that they are, for the most part, artifacts of an incomplete fossil record. Thus, in the example already cited of numerous closely related species appearing simultaneously, he essentially claims that the record presents us with a *non*pattern. The record, he says, is condensed, not enough time being recorded in the sediments to allow us to see the temporal succession of forms that would be predicted "under his theory."

Nonetheless there are two patterns that Darwin denies—or, at least in one case, is at pains to explain away—that turn out to have been "real" and to have required positive explanation rather than excuses. The first is the absence of "transitional" forms: Why, in other words, are species discrete? Though Darwin blames the inadequacies of the fossil record for the absence of intermediates between fossil species (in Chapter 9), he has no similar convenient excuse for, and thus must squarely confront, the discreteness of species in the modern biota. This he does in part by cleverly invoking a somewhat nebulous model of geographic divergence, and claiming, at one point, that "(I) believe that species come to be tolerably well-defined objects. . . " (p. 177, Chapter 6). However, his major explanation is that extinction—through the agency of competition, where superior descendant species tend to wipe out their inferior forebears—is the main cause of gaps between closely related contemporaneous species.

Darwin has been criticized persistently on this very issue, ever since initial publication of the *Origin*. Naturalists, perhaps beginning with the German Moritz Wagner, and the English biologist G. J. Romanes, repeatedly pointed to the discontinuity between species and the seeming importance, not just of geographic variation, but of the isolation caused by geographic disjunction as an indispensable aspect of the evolutionary process. In other words, these and other early critics saw a need for a *positive* mechanism that would account for discontinuity among species.[22]

Stressing connectedness and continuity also led Darwin to deny another important empirical pattern in the fossil record: *stasis*. Philosopher David Hull, in his *Darwin and His Critics* (1973), reprints all the early reviews of the first edition of Darwin's *Origin*, including some four or five by paleontologists. *All* of them remark on the absence of any truly significant discussion of the well-known fact (to paleontologists) that *once a species appears in the fossil record, though it may exhibit normal variation both within local populations and geographically, little net evolutionary change tends to accumulate throughout its entire duration.* This duration is now known to be typically 5 to 10 million years, at least in the case of marine invertebrate species. But Darwin did acknowledge that some species change more slowly than others (see above quote and note 20); he responded more fully to his critics on just this point in his sixth edition. Yet stasis remained an embarrassment to him, because the actions of natural selection should on first principle act to perfect and thus by definition modify adaptations were the environment to remain stable. Either that or it would modify species to keep fitting what mid-nineteenth-century geologists had been pointing out: that the earth, including its climate, is continually changing.

For the most part, Darwin simply denied the reality of stasis. Instead, he chose once again to blame the nonprevalence of "incessantly graded series" in the fossil record on a poor record: poor preservation, lack of time documented in sediments, and lack of paleontological collecting and analytic experience. But none of these reasons could explain away the problem for later generations. Just as discontinuity among living species became the pattern to be explained for Dobzhansky and Mayr, stasis plus discontinuties between fossil species became the pattern to be explained for Eldredge and Gould. Darwin's main concern was to convince the world that life had evolved. After all, he could not have been expected to have explained absolutely everything; he got an awful lot. To him, discontinuity and stasis were two patterns that were inconvenient, if not wholly antithetical to establishing his views.

Darwin on Natural Selection: Chapters 1 + 2 + 3 = Chapter 4

Physicists have never been much good at understanding natural selection. Darwin was mortified when John Herschel, an outstanding physical scientist from mid-nineteenth-century England, pronounced natural selection as the "law of higgledy-piggledy." Closer to the modern era, the late physicist-turned-philosopher Karl Popper condemned evolutionary biology as "unscientific," in large part because natural selection made no sense to him as a scientific principle (He had other reasons as well; see note 13 in this chapter.) But he recanted this viewpoint, virtually on his deathbed.

Why do physicists, who have the reputation of being among the best and the brightest, have such a hard time with the simple notion of natural selection? For simple it is:

> As many more individuals of each species are born than can possibly survive; and as, consequently, there is a frequently recurring Struggle for Existence, it follows that any being, if it vary however slightly in any manner profitable to itself, under the complex and sometimes varying conditions of life, will have a better chance of surviving, and thus be *naturally selected*." (*Origin*, 1859, p. 5)

In slightly expanded form, Darwin and Wallace saw that organisms vary. (No two are exactly alike, except possibly "identical" twins.) They also recognized that organisms tend to resemble their parents: Variation is heritable. More offspring are produced each generation than can possibly survive and reproduce. Otherwise, the world would soon be overrun by individuals of just one species; Darwin calls this concept the *doctrine of Malthus*. Because this is so, on average those organisms whose heritable features best suit them for making a living will survive and eventually reproduce. Obviously, the recipe for success will be handed down to the next generation as well. Things may change when new heritable variants appear, or especially when environments change. Thus will the formula for successful living be changed, and new variants will be favored over all.

That's it. The concept is definitely simple enough. This description of natural selection may be a bit longer than the elegantly brief $F = MA$. Conceptually, however, it is hardly more complicated.

Biologists tend to attribute the difficulty physicists typically have with natural selection to their inability to think in terms of populations, genetic variation, and statistical effects (that is, the "on average" component of natural selection as spelled out in the preceding paragraph).[23] I believe the real reason lies elsewhere: Physicists are accustomed to thinking of natural phenomena in terms of particles and forces. For that matter, so are chemists as well as geologists, going all the way back to James Hutton at least. They look at two sets of "things": the material entities of the universe, and the forces that act upon them. They think about energy. Take the gas laws (which are, after all, statistical): There are regular relations between volume, pressure, and temperature of a collection of gas molecules trapped in a box. Molecules collectively apply more pressure to the walls of the box dependent on the amount of energy, such as heat, applied to the system.

Natural selection is very different from any such "law" in physics. Nothing in the post-Darwinian era of genetics and molecular biology has changed the original formulation to explicitly include "forces," in the sense of energy flow. Particles, perhaps, enter into the formulation. Genes are particles of a sort. But here we must be careful, as what is "selected" is the information they carry rather than individual genes themselves.[24] But energy? No. In fact, evolutionary biologists routinely speak of natural selection itself as a "force." It is not. It is an information filtration process. Thus it was nearly a century ahead of its time, for not until the early days of "information theory" did information emerge as a separate, though in many ways a parallel, world to the far more familiar realm of particles and physical forces.

Darwin's first four chapters in the *Origin*, as I have already remarked, are a series of discussions of patterns that lead up to the overwhelming conclusion that nature has her own version of the process of human, or artificial, selection: selective breeding, which has been in use since the Neolithic Age (that is, for at least 10,000 years).[25] Although as a callow undergraduate first tackling the *Origin* I was much disappointed to find myself immersed right off the bat in a discussion of pigeon breeding.[26] Darwin's strategy of leading his reader down an inductive garden path—while always admitting along the way that he is leading to the notion of natural selection, is brilliant. First, he shows what many people already must have known to be true: that domestic animals and plants exhibit patterns of variation that are heritable. His main point in the first chapter is a grand analogy: The "varieties" of domestic pigeons (or dogs or virtually any other domesticated species) are analogous to the varieties found in nature—and natural varieties are sometimes difficult to distinguish from separate species: his basic argument of continuity.

Chapter 2 of the *Origin* reviews comparable patterns of heritable variation in nature. Though Darwin openly laments the lack of detailed information, he nonetheless finds much evidence to support his generalization, his *pattern*, that most species display variation in the natural world. Here, again, his argument of continuity in pattern is coupled with his further goal: to establish that the sort of heritable variation routinely shaped by selective breeders is also very much to be found in the natural world.

Darwin then turns, in Chapter 3, to the critical data supporting his contention that the Malthusian doctrine applies to the natural world just as it does to humanity: "many more individuals of each species are born than can possibly survive." Here pattern arises not so much from repeated observations of hauntingly similar cases, but from a form of mathematical necessity. Left unchecked, populations grow logarithmically. Darwin's hypothetical illustration using elephants is somewhat fanciful but very much to the point. Their slow rate of reproduction only enhances his case:

> There is no exception to the rule that every organic being naturally increases at so high a rate, that if not destroyed, the earth would soon be covered by the progeny of a single pair. Even slow-breeding man has doubled in twenty-five years, and at this rate, in a few thousand years, there would literally not be standing room for his progeny.[27] Linnaeus has calculated that if an annual plant produced only two seeds—and there is no plant so unproductive as this—and their seedlings next year produced two, and so on, then in twenty years there would be a million plants. The elephant is reckoned to be the slowest breeder of all known animals, and I have taken some pains to estimate its probable minimal rate of natural increase: it will be under the mark to assume that it breeds when thirty years old, and goes on breeding till ninety years old, bringing forth three pair of young in this interval; if this be so, at the end of the fifth century there would be alive fifteen million elephants, descended from the first pair. (*Origin*, p. 64)

Clearly, the world is not "standing room only" in anything—certainly not in elephants, nor even in humans, at least for the time being. Darwin's claim that the empirical data for prodigious increases in population numbers constitutes "better evidence than mere theoretical calculations" (p. 64) is at best a debatable contention, given the logical necessity and sheer force of the "theoretical contentions" of Malthusian calculations. After Darwin reviews some such examples, he turns to the thorny issue of what in nature actually controls the numbers of a species. This discussion is probably the earliest cogent one of what still proves to be a contentious issue.[28] Darwin stresses food supply, climate, disease, and predation as the prime generic suspects in the control of population size.

The issue of population size control is critical. For here lies the trickiest part of the analogy between *conscious* human artificial selection and the *unconscious* weeding process that otherwise occurs naturally in the wild. In Darwin's imagery, nature does not *consciously prefer* one variant over another. Rather, it's all what comes out in the wash, in the "struggle for existence" that often spells competition within populations for resources. Energy resources (food for animals, sunlight and nutrients for plants) are generally, and correctly, held to be finite. Hence they are a limiting factor on a local population. Some members of a population will be more efficient than others in exploiting that resource. On average, those more efficient at finding and consuming such limited energy supplies will survive (as Darwin put it) and be more likely to reproduce (as modern evolutionary biologists are more prone to stress). Though frequently characterized as "competition," it really is a matter of which variants are best able to cope with the situation. That situation could involve finding something to eat: Animals rely ultimately on the population sizes of their prey, which are themselves limited. Or it could involve surviving climatic factors (for example, extreme heat or cold), disease, or predation.

Thus, when we arrive at the *Origin's* Chapter 4, "Natural Selection," the argument becomes one of integrating the two components: heritable variation and the struggle for existence, the latter with a built-in Malthusian dynamic. Darwin attempted to integrate them into a "what works better than what" formulation that he called *natural selection. Natural* as opposed to *artificial* makes abundant sense. Still, I have often wished that Darwin had not chosen the term *selection,* with its purposive, intention-laden baggage. Nature doesn't care who survives. The process is just a matter of what works better than what. Given finite resources, and in modern parlance, the genes that specify the more efficient, successful heritable variants will be the ones that are disproportionately passed along to the next succeeding generation. What we have here is not so much *selection* as *filtration* (though it may not sound as good). Natural filtration of heritable information.

Darwin himself well remembered the spot on the road when the idea of natural selection first came to him, like a thunderclap, a genuine "eureka." Here was a piece of creativity that came from putting disparate patterns together: the wealth of evidence of connectivity; the ability of humans to mold the heritable features of domesticated species; until then missing ingredient: the realization that the Malthusian principle of geometric population expansion is routinely denied. Thus resources and other limiting factors themselves will serve to bias the representation of heritable variation in the next generation.

So I am convinced that Darwin started with his argument on natural selection, not because he felt he needed a mechanism per se to be convincing, but because he felt that the record of breeders was a de facto record of evolution, albeit human induced. To show that an analogous process occurs in the natural world is ipso facto to show that evolution takes place in the wild. It was patterns of biased inheritance, filtration of heritable features, in both systems that seemed to Darwin the very best evidence that "descent with modification" is a fact of nature.

As I have already remarked, Lamarck too had a "mechanism," one based on the inheritance of acquired characters. This concept was one that Darwin himself partially embraced in later editions of the *Origin* in addition to his core idea of natural selection. But Darwin succeeded where Lamarck and others had failed to establish the idea of evolution because his patterns were so convincing. Indeed, philosopher David Hull[29] has argued convincingly that the scientific world all but unanimously embraced the idea of evolution upon the publication of the *Origin*. It was natural selection, not "descent with modification," that proved to be contentious and far from universally accepted, possibly all the way up to the work of geneticists Ronald Fisher, J. B. S. Haldane, and Sewall Wright in the 1920s and 1930s.

But note Darwin's choices for the "limiting factors" on population size—hence those factors responsible for this "natural selection." They are exclusively *economic* in nature. In other words, they pertain to matter-energy transfer processes concerned with the development and continued existence of the organism itself: its search, consumption, and utilization of energy resources and nutrients; its survival against climatic effects, disease, predation, and so on. Though natural selection is a filter of genetic information, most of that information pertains to the economic lives of organisms. In that important sense, Darwin had actually forged a conceptual link between organisms and the world of entities and forces, the physical world more familiar to nineteenth-century science.

Indeed, in his fourth chapter in the *Origin*, Darwin made a fundamental distinction between *natural* selection and what he termed, in contrast, *sexual* selection. The distinction is valid. Although its implications may remain to be fully absorbed even by modern biology, nonetheless it provides a glimmer of understanding about the true nature of the connection between the evolution of life, on the one hand, and the world of (nonliving) entities and the physicochemistry of forces. For in drawing a distinction between two forms of selection, Darwin saw that organisms live in two distinct realms. One involves the economic issues of existence, the other solely concerns itself with the business of reproduction.

Darwin defined sexual selection simply as "a struggle between the males for possession of the females" (*Origin*, Ch. 4, p. 88). Much more thoroughly and convincingly, in his *Descent of Man* (1871, p. 256), Darwin expanded and generalized this earlier definition of sexual selection, now seeing it arising from the "advantage which certain individuals have over other individuals of the same sex and species, in exclusive relation to reproduction." Selection is often characterized (in post-Darwinian times) as "differential reproductive success." For this, there are two general causes. First, some organisms thrive relatively better than others in the wild, meaning that differential *economic* success begets, on average, differential *reproductive* success. Those organisms better able to find food and fend off the elements will tend to survive and reproduce—natural selection. Second, some organisms may just be better at getting mates than others. Indeed, the complex mating behaviors and morphologies of many species suggested to Darwin that competition for mates might well drive the evolutionary modifications of plumages, calls, dances, and other traits. Development of such features may be geared, not to making a living per se, but simply to attract mates. Thus Darwin's notion of sexual selection complements natural selection. In a finite world of resources and potential mates, those best equipped for the job will tend to carry the day. The heritable ingredients for success will be handed down differentially to the next succeeding generation.

Until recently biologists have tended to ignore the distinction between natural and sexual selection that Darwin first drew in 1859. But it underscores the connection that natural selection has with the physical world of matter-energy transfer. For in drawing the distinction between two sets of causes, each of which bias transmission of genetic information from one generation to the other, Darwin is effectively demonstrating that selection in any form is different from a typical physical process familiar in conventional science. As we've seen, selection is about the transfer of information (even though Darwin's own speculations on heredity have proved to be completely false). Selection it is not a statement about matter-energy transfer, or any other form of physical interaction. Rather, it is a statement about what happens to heritable information *as a consequence* either of such physical interactions or competition for reproductive success.

Richard Dawkins has been especially eloquent on the subject of natural selection as an information-filtering system. With the advent of modern understanding of the molecular biology of the gene, and the continued development of information theory, biologists have become increasingly comfortable seeing selection as a process of differential information transfer. But with this welcome vision has also come an accentuation of themes that serve to exacerbate the non-connectedness of the living world with the forces and entities of the physical world.

For example, Dawkins[30] is well known for his metaphor of the "selfish gene," where the notion of organisms competing for reproductive success is projected downward (in "good" reductive fashion) to the level of the gene. Genes, in Dawkins's vision, are competing for representation in the next generation. And though such a formulation seems like a logical extension downward of Darwin's notion of *sexual* selection, it actually subsumes true natural selection as well. For example, competition for economic resources such as food and water is typically seen as merely a manifestation of competition for reproductive success—an inversion of the original Darwinian paradigm. If Darwin saw competition for resources leading to differential reproductive success, modern biologists like Dawkins tend to see competition for resources as a manifestation of competition for reproductive success, itself an urge traceable to the genes themselves.[31]

Echoing some of Darwin's strongest themes and denials, modern evolutionary biology emphasizes genes and biotic interactions at the expense of explicit links to the physical world. Ecology plays a muted role in evolutionary theory. And patterns in the history of life—patterns strongly linked with the physical history of the earth—still await full integration with evolutionary theory. That natural selection is not the same kind of principle as, say, any of the "laws" of physics and chemistry should by now be obvious. That such a conclusion neither renders natural selection "wrong" or "unscientific" should also be obvious. All subsequent mathematical, theoretical, experimental, and field observations reveal what first became obvious from Darwin's first four chapters of the *Origin:* Natural selection is an ineluctable "law" of the natural world. But it is not the kind of law that many evolutionary biologists (ultra-Darwinians, in my language) have tried to force it to be, as when they cast organisms, and especially the organisms' genes, in "active" roles, competing for reproductive success. Organisms do interact — with members of the same species, with members of other species, and directly with the physical world. Natural selection is the fallout, the record of history, from those interactions. But it is the interactions themselves that constitute the very real connections between biological systems and the rest of the physical world.

CHAPTER 4

Evolution of the Earth

James Hutton's legacy has been the vista of deep geological time, a virtually limitless span of cycles of uplift and erosion punctuated by occasional volcanic and seismic events. Because he could see "no vestige of a beginning, no prospect of an end," Hutton was able to meld skillfully a view of the dynamics of the earth with a strategy for interpreting its history. He saw clearly the energy sources that drive those processes: the endogenous heat flowing from the earth's interior, now known to be a product of radioactive decay; gravity; and the winds, currents, and other effects of solar radiation, or *exogenous energy*. But Hutton's vista had no place for strong linearity—no sense of direction. Nor did it have room for any notion that the earth's single history could be read as a progression of interrelated events. History, yes, but evolution, not really.

Hutton's successors immediately began to tease apart the study of earth processes—volcanism, earthquakes, sedimentation—processes understood to be forever ongoing during the course of earth's history—from the analysis of the course of events that had shaped the earth through geological time. Lyell was the great exception. Forever mindful of Hutton's great uniformitarian principle, never forgetting that "the present is the key to the past," he was forever committed to a steady-statism closely allied to Hutton's cyclicity.

Nevertheless, a strong sense of linearity and direction quickly emerged. It did so primarily through the efforts of the stratigraphers and paleontologists whose first collective effort revolved around the construction of the geological timescale. Baron Cuvier and his colleagues in France, plus William Smith and the many English geologists who quickly followed him, became adept at deciphering local sequences of layered rocks. By tracing horizons laterally, they were able to piece together the details of the entire geological column, or at least that part that represents what we now realize is the last half billion years of the 4.6 billion years of geological time.

These early efforts focused on relatively undisturbed sedimentary rocks, mostly laden with fossils. And fossils were the critical factor to the historical geologists. As we saw in Chapter 2, the fossil content of strata changes as one climbs up and down a geological section. This variation enables the paleontologically informed stratigrapher to distinguish the strata in the first place and to recognize strata of roughly equivalent ages elsewhere. That the same or similar fossils always appear in the same order wherever they are encountered around the world was enough to impart a strong sense of linearity to earth history.

But the clincher, of course, was Darwin's utter victory in convincing the scientific world of the mid-nineteenth century that life had evolved. He, more than any of his contemporaries, argued successfully for an age of the earth in the hundreds of millions of years. Thus he added significantly, yet in more concrete terms, to Hutton's unspecified boundless vista of time. But Darwin also explained why the fossils of successive strata change through time: Life has evolved, and the changing fauna and floras leave their traces in sediments deposited progressively through time.

Yet a volcano is a volcano, whether it erupts in A.D. 79, as did Vesuvius, or during the Cretaceous Period, as did the dominant hill of Montréal. There is little about uplift, erosion, or any other readily observable geological process comparable to biological evolution. Repeated historical patterns—whether signs of volcanism, or the sorts of unconformities so crucial to Hutton's thinking, or indeed the repeated patterns of extinction and evolution[1] in biological systems—are important primarily for what they reveal about the general nature of historical process, be they geological processes or biological evolutionary processes. Despite the early successes of geologists in framing the outlines of the geological timescale, despite Rutherford's application of radioactivity to plugging in real-time ("absolute") dates to the divisions of geological time at the end of the century, nothing like an actual evolutionary theory of the earth was available even by the dawn of the twentieth century.

Philosophers and scientists alike are fond of the idea that an evolutionary theory pertaining to any sort of system is unlikely to be established without a convincing *mechanism*, a well-articulated and corroborated *process* of change. As we saw in the preceding chapter, Darwin is still commonly supposed to have succeeded where all others before him had failed in establishing the credibility of evolution because he supplied a process that could plausibly account for the history of life: natural selection. Yet natural selection far more than the basic notion of evolution itself has provoked debate in the nearly century and a half since Darwin. The contention involves whether pattern is preeminent over process in establishing the credibility of an evolutionary theory. Much this same debate has swirled around the geological evolutionary theories of continental drift and plate tectonics. Why was Alfred Wegener's (1880–1930) theory of continental drift resisted so strenuously for so long, only to emerge triumphant in the reconstituted guise of "plate tectonics" in the 1960s?

Prior to the twentieth century, there were various theories of the origin of the earth and historical development of its basic features. Three basic patterns of the earth's surface attracted particular attention: the differentiation of continents and ocean basins; the localized concentrations, on continents, of very thick sequences of sediments and associated igneous and metamorphic rocks, often associated with continental margins; and the linear belts of deformed and uplifted rocks, or mountain ranges.

Around the turn of the century, the prevailing view of the origin and history of the basic structure of the earth was embodied in the work of the Austrian Eduard Suess (1831–1914). His views were expressed in *Das Antlitz der Erde* (*The Face of the Earth*), which was originally published in the 1890s, and translated into English in five volumes during the first decade of the twentieth century.[2] Suess believed that the earth's mountainous fold belts, mainly restricted to the margins of continents, were contractions of a shrinking earth, which was cooling from a molten origin. The discovery of radioactivity and the quickly formed supposition that heat flow emanating from the earth's interior is the product of nuclear decay, dealt a lethal blow to Suess's theory of the earth's molten origin, and also to his ideas on the origin of mountain belts. Favored now is a theory by American geologist T. C. Chamberlain that there was a cold, accretionary origin of the earth through the assemblage of many smaller bodies. This theory is known as Chamberlain's *planetesimal hypothesis*.

Other Suess' notions have fared better. It was Suess who coined the terms *sial* and *sima*. Still in general use, sial refers to rocks rich in silicon and aluminum: the granitic rocks forming the cores of continents. Sial is less dense than sima, the high silicon and magnesium rocks of the oceanic crusts. Geodesists, who study the contours of the earth, had discovered in the nineteenth century that mountain ranges have less mass than their sheer bulk would seem to imply, leading to the theory of isotasy, which sees lighter, continental rocks riding above a layer of denser rock. The higher an elevation, such as a mountain range, the deeper its roots are embedded in the denser sima below. Sial "floats" on sima, just as ice cubes float on water. Picture an ice cube: The larger it is, the deeper it projects down into the watery medium, but also the higher it sits above the water's surface. All this was evident to Suess, and in this respect his basic concepts remain.

It was Suess, too, who coined the term *Gondwanaland* for an ancient continent consisting of parts of Africa, India, and Madagascar. (Later it was shown to have also included Australia, South America, and Antarctica.) Here we confront historical pattern for the first time in

Suess's theory. The existence of Gondwanaland was predicated on the existence of very similar plant and animal fossils found on each of the now–widely separated continental land masses. Unlike Wegener, though, Suess hypothesized sunken land bridges, portions of continents that have foundered and disappeared below the waves, the result of an earth contracting as it cools.

But then there is the problem of the massively thick sequences of sedimentary and volcanic rock exposed in the world's great mountain ranges: the Andes, Rockies, Alps, and Himalayas. The invertebrate fossils in these sequences demonstrate clearly that the sediments were deposited in ancient seaways. The marine sediments over most of the continental interiors typically form a much thinner veneer. Why exactly are the piles of marine sediments so much thicker in mountain ranges than they are elsewhere on the earth's continents? Long before Suess, geologists had struggled to explain how long linear depressions would form on continental margins. American paleontologist James Hall drafted his theory of such *geosynclines* in 1859, noting that the crust at the edge of continents must be sinking at near the same rate that the thick layers of sediment were accumulating. But what causes the crust to warp? The bobbing effect of isostasy would tend to counteract downward pressure of the sheer weight of sediments. Also, sediments would quickly fill the basins, and their weight alone would hardly suffice to sink the basins still farther below sea level. Suess saw the origins of such deep sedimentary basins as evidence of foundering of the crust, part of the dynamics of his shrinking earth. And where were all those sediments coming from, anyway? Geologists soon discovered that at least *some* of the sediments in mountain ranges like the Appalachians seemed to have come from the *east*, where only the watery depths of the Atlantic now stand.

It had become reasonably clear that continents are older in their centers and younger towards their edges, and that they grow by a form of accretion, the addition of linear, sediment-gorged troughs subsequently deformed into a mountain range, one after another. Much later, it also became apparent to geologists like Marshall Kay that modern-day island arcs (such as the Aleutians, Japan, or the Philippines), with their associated earthquakes, volcanoes, shallow seas between them and the mainland, and deep-sea trenches on the ocean side, are modern-day examples of these linear geosynclinal belts in the making. It seemed to Kay that the thick sedimentary deposits accumulating between the islands arcs and the continental mainland, as well as the sediments and volcanic outpourings cumulating seaward in the deep trenches, were precisely the same as sequences familiar to him from many of the world's deformed mountain belts.[3]

Nevertheless, as difficult as it was to understand how deep basins formed, were filled with sediment, and eventually uplifted into vast mountain ranges, most geologists insisted that the earth is essentially stable. (Most, that is, save the Wegenerian minority.) Continents might well founder and sink, and sometimes rise again, but they could not move around vis-à-vis one another. There was no satisfactory mechanism for it. Besides, the crust was considered far too brittle to allow the sialic continents to plow around a sea of sima. The crust of the earth, for all its earthquakes and volcanism, is essentially stable, or so I learned as an undergraduate geology major at Columbia College as late as the early 1960s. However the Columbia faculty was even then heavily engaged in the plate tectonics revolution, so that by the time I finished graduate school (1969), everyone "knew" that continents had indeed changed their positions with respect to one another extensively over the course of geological time. To doubt that "fact" in 1969 was as silly as saying only five or six years earlier that it was so.

Why were Wegener and others so unsuccessful in establishing what came in a rush in the mid-1960s? True, when Wegener wrote, he lacked some vital information and misinterpreted other data. Still, his proposed mechanism for drift was not all that different from later views adopted under plate tectonics. And his main source of evidence, his main *patterns*, were the same as those cited now in support of plate tectonics, (though, as we shall see, additional geophysical patterns were needed before the revolution could become complete.) Wegener, in a sense, was like Lamarck: both men had plausible mechanisms. (For their time, Wegener's mechanism was actually far more accurate than Lamarck's use-and-disuse mechanism for the evolution of adaptations.) Both emphasized patterns that, in retrospect, are perfectly valid signals of evolution (the *scala naturae* for Lamarck; a host of patterns for Wegener). Yet neither managed to convince the majority of their contemporaries. Why? Not because their theories of mechanisms were so faulty, but because their choice of pattern was incomplete and insufficient to do the job. It was up to Darwin to fashion his intricate web of argument and pattern, and up to an international community of geologists and especially geophysicists to compile their own historical patterns, before both of these theories of evolution could be openly and enthusiastically embraced by the scientific world.

What Did Alfred Wegener Say?

By his own account, Alfred Wegener first came to his hypothesis of continental drift in 1910 "when considering the map of the world, under the direct impression produced by the congruence of the coastlines on either side of the Atlantic." This idea, he said, he at first disregarded "as improbable."[4] As he himself recounted, he was not the first to draw attention to the congruence of Atlantic coastlines. Nor was he the first to suggest, on that basis as well as on a multitude of other patterns, that the continents change position with respect to one another throughout the course of geological time. Yet it was Wegener who sparked the debate, provoking the negative reaction that dominated geological thinking, especially in the United States. And it was also Wegener who sowed the seeds of change by convincing a few of his younger colleagues of the basic truth of his ideas. He set the stage for the ultimate emergence of plate tectonics in the 1960s. That new direction in geological thinking was revolutionary with respect to the intransigent opposition his ideas had provoked for 40 years.

Wegener organized his book into a series of "arguments"—not unlike Darwin's own characterization of the *Origin* as "one long argument." Wegener himself was a meteorologist, with training in astronomy and physics. He was a relative outsider to the already-professionalized world of geology and the newly emerging specialty of geophysics. Turning first to the world he perhaps knew best, Wegener not unreasonably supposed that a direct demonstration, an actual measurement, of yearly movement going on right now would be the best, most convincing evidence that continents change their positions vis-à-vis one another. And Wegener claimed to have these numbers in abundance for many different crustal blocks.

The most important of these crustal movements was Wegener's claim that Greenland had been drifting westward at a variable rate of between 9 and 32 meters a year. (Greenland had been a boyhood passion of Wegener, and he ultimately perished there on his third expedition.) The rate of Greenland's westward drift was calculated by Wegener based on a series of longitudinal observations relative to the Greenwich meridian. Though Wegener was aware that these measurements were fraught with error, he nonetheless maintained that the calculated displacement exceeded the error estimate by a considerable margin. Thus he concluded: *"The result is therefore proof of a displacement of Greenland that is still in progress"*[5] [italics are Wegener's].

Later measurements failed to confirm both Wegener's data and his conclusions, and this became a source of much of the skepticism of his views. Lasers have subsequently allowed far more precise measurements. They have in fact detected rates of drift (seafloor spreading) much slower than Wegener claimed—generally on the order of 1 to 2 centimeters per year. For example, direct measurements on Iceland—a part of the Mid-Atlantic Ridge from which new seafloor is being generated and spreading both east and west—reveal a separation of about 1 to 2 centimeters a year. Rather like Darwin, who grandly overestimated the age of the Weald as a landform in his enthusiasm and desire to establish the credibility of deep geological time, Wegener had vastly overestimated the rate of continental displacement.

Wegener did not fare much better with his next selection of arguments, those based on geophysics. Here, Wegener points to the bimodal distribution of elevations of the earth's crust: the very marked differences between the elevations of the continents and the oceans. In this lies the proof, he says, that the crust underlying the continents and oceans is indeed very different—with the light sial forming the continents and the denser sima composing the oceanic crust. He argues that the lighter sial of the continents floats on the denser layer of sima below. Or, put another way, the oceans lack an upper layer of sial.

For drift to occur, for the sialic continents to be able to move around, they must be able to plow through, or be carried along by, a plastic medium. Wegener argues that the very phenomenon of isostasy—particularly his version of it, which sees blocks of sialic crust floating like variously sized ice cubes extending down into a lower layer of denser sima—supports his notion of continental drift. Geologists and geodesists had already documented the phenomenon of *isostatic rebound:* northerly reaches such as New England and Scandinavia, until 12,000 years ago covered with as much as a half-mile-thick layer of ice, are still rebounding, or rising, now that the heavy layer of ice has been removed. Isostatic rebound, to Wegener, was proof positive of a plasticity to the inner layers of the earth, which needs to be there if the sialic continents are to be able to move about the earth's surface.

Much has been made of the resistance that Wegener met in the mid-1920s, especially at meetings of the British Association for the Advancement of Science (1922), the Royal Geographical Society (1923), and later the famous 1926 New York meeting of the American Association of Petroleum Geologists. Wegener did have his defenders at each of these venues. (He himself was present in New York, though apparently most of the pro-Wegnerian commentary was supplied by the meeting's chairman, the Dutch geologist W. A. J. M. van Waterschoot van der Gracht.) But Wegener's opponents were many. They considered all of his arguments and derided many of his interpretations of geological, paleontological, and biological patterns, even his matchups of the Atlantic coastlines. By far the most damning criticism was of Wegener's geophysical arguments. The leading figure in geophysics in those days was the British scientist Harold Jeffreys. He spoke for the clear majority of his colleagues when he pronounced the sima far too weak and brittle, entirely too nonplastic, to allow the sort of movements Wegener envisioned.[6]

Here again the central contention of *mechanism* rears its head. In this instance, it is not so much the lack of one, but the implausibility of Wegener's notions. For not only did geophysicists deny the sort of plasticity required, they were quite skeptical, and rightly so, about the basic forces that Wegener postulated were behind continental drift. It is important to understand that these meetings, plus sundry scathing reviews, which served to discredit continental drift for 40 years, took place before Wegener had modified his views on mechanism. In his third edition (1922), the first to be read widely, Wegener was still insisting that tidal forces (essentially the gravitational effect of the moon on the earth) and what he called the pole-flight force (*Eotvos*) — the tendency of the continents to move away from the poles through the centrifugal force of the earth's rotation — produced continental movement. Jeffreys pronounced them as far too weak to do so.

In his last edition Wegener admits (p. 167) that "the Newton of drift theory has not yet appeared." But he still retains and defends these tidal and pole-flight forces as necessary and sufficient for continental movement. (However, he thought the pole-flight force could not produce folded mountain ranges, as the American geologist Frank Taylor had claimed.) Nowadays, tidal forces and pole-flight forces are all-but-forgotten historical curiosities. They play no role whatever in modern geophysical theory.

Wegener adds a critical few paragraphs on convection of the sima in the last edition of *The Origin of Continents and Oceans*. The modern theory of plate tectonics sees convection as central to the process of generating new crustal material. As new crust is generated, at midocean ridges, plates spread apart, carrying continents along with them. One can only wonder what the effect Wegener's ideas on convection might have had if he had articulated them in the third edition. They were so close in their embryonic way to the basic mechanistic notions of modern plate tectonics.

Wegener's Historical Patterns: Geology, Paleontology, and Biology

To the general fit of the Atlantic coastlines, Wegener had many more patterns to add. Delving into the geological and paleontological literature, he was quite naturally drawn to the plethora of examples of intercontinental matchups of the southern hemisphere, patterns that had led Edward Suess earlier to formulate the concept of Gondwanaland. The difference between Wegener and Suess on the subject was simple, yet profound. For Suess the old land bridges had sunk below the waves to become ocean basin. Wegener, convinced that lighter sialic rocks could not so founder, instead attributed the similarities to common history: Gondwanaland was a single megacontinent now torn asunder and rifted apart by the forces of continental drift.

Much as Darwin before him had turned to patterns in the geographical distribution of plants and animals to argue that life must have had a long evolutionary history, Wegener thought that biogeographic patterns also constituted an "argument" for continental drift. Close relatives separated by large intervening spaces seemed to him prima facie evidence that at one time the regions had been closer together. For example, Wegener cites biological literature on the distribution of earthworms that suggests that the continents in the southern hemisphere had to have been in close conjunction with one another. As we shall see, the paleontological evidence also offers especially compelling support for Wegener's interpretation of Gondwana. But such biological patterns (that is, those involving living organisms) were easily dismissed or interpreted differently, and they hardly constituted the most formidable of Wegener's "arguments."

It is ironic that the very same features commonly listed today as "proof positive" of the rifting apart of a single ancient landmass were among the many patterns Wegener himself listed, which largely fell on deaf ears in the 1920s. And yet the evidence was hardly lacking. By the turn of the century, steamships had taken to stopping in Antarctica to lay on coal supplies. Coal in Antarctica! Geologists and paleontologists had long since realized that coal deposits generally formed from compressed layers of partly decayed plant matter accumulating in the bottoms of swamps. And such swamps are generally seen as tropical or at least semitropical in origin. For example, Glossopteris, an Upper Paleozoic plant, had been described from such far-flung reaches as India, South America, southern Africa—and even Antarctica.

Nor did all the evidence of such rifting involve plant life. The two-foot long Mesosaurus is a particularly compelling fossil. A graceful little reptile with a long snout and jaws equipped with many spikey, fish-catching teeth, it had turned up in Argentina and South Africa (and is still apt to turn up on Graduate Record Exams in geology). The catch was that Mesosaurus was known only from lake sediments. It was a freshwater animal clearly not able to traverse the vast waters of the South Atlantic Ocean.

Wegener made the most of such data: the individual cases that added up to marked patterns of historical resemblance among the continents of the southern hemisphere. (India, too, was once part of the Southern Hemisphere. It has subsequently traveled north over the equator, eventually colliding with the Asian plate to produce the Himalayan fold belt.) Drawing on the literature, Wegener matched mountain chains in South America and Africa, and tabulated a series of other similarities in the igneous, metamorphic, and sedimentary rocks of these regions that are now widely separated.

The grandest pattern of all that linked these Gondwanan continents was the same *historical sequence* of geological and biological history found everywhere. With evident satisfaction, Wegener recounts a detailed list of some 15 events and phenomena documented by the South African geologist Alexander DuToit in both Argentina and South Africa.[7] Starting with the Precambrian and early Paleozoic, DuToit's summary of geological deformation and paleontological congruity, in exactly the same order on both sides of the Atlantic, seems as compelling a summary of pattern as could be imagined.

One relatively minor example has special meaning for me. DuToit briefly mentions that the Devonian faunas of southern South America are quite different from the Devonian of northern South America and northern Africa, not to mention North America, Europe, and Asia. My students and I have spent 25 years focusing on the Upper Silurian– Middle Devonian trilobites of the Andes (primarily in Bolivia, but also in southern Peru and northern Argentina), and the eastern basins of Argentina, Uruguay, and Brazil, plus the Falkland (also known as the Malvinas) Islands, South Africa, and, to a limited extent, Antarctica. Though trilobites of many different higher taxonomic affinities are present throughout these sequences, the faunas are everywhere dominated by species belonging to a single family, the Calmoniidae. The calmoniids are the last great evolutionary radiation in trilobite history, and their anatomical diversity is remarkable. They are found nowhere else in the world. We now know them to have lived in shallow, very cold marine waters: The South Pole in the Devonian was located a scant 100 miles north of present-day Cape Town. Put the continents back together, and you get a single, high-latitude region of flooded continents, divorced geographically and ecologically from the rest of the Devonian world.

Though my colleagues and I have described literally dozens of new species, and many new genera, the general facts of Devonian trilobite diversity were well known to DuToit, primarily through the work of F. R. Cowper Reed, who wrote a section of DuToit's book. The point is that such patterns should be no more intrinsically compelling now than they were in the 1920s. Yet they are cited today as critical evidence of drift, whereas they were routinely ignored in the 1920s.

Similarly, when my professor, E. H. Colbert, a vertebrate paleontologist specializing in the reptilian faunas of the Triassic Period, went off to Antarctica in the 1960s with South African paleontologist James Kitching, they detailed striking similarities in the stratigraphy of Antarctica and the famous Karroo beds of South Africa. Colbert gained some notoriety by describing the lower jaw of the Triassic reptile *Lystrosaurus*, previously known only from the Karroo of South Africa. Widely hailed as a final nail in the coffin of stabilist theory, the paper was published just as the more sophisticated plate tectonics version of continental drift was catching on in the mid- to late 1960s. Yet Colbert would have been the first to acknowledge that *Lystrosaurus* really carried no more weight than the much older pattern of *Mesosaurus*.

But how are patterns, so easily interpreted once the paradigm is shifted, so easily dismissed or explained away even as the core ideas are initially developed and proposed? Anthony Hallam[8] is surely correct when he suggests that the absence of a plausible mechanism itself cannot be the whole story. After all, Hallam points out, continental glaciation theory, first proposed by Louis Agassiz in the nineteenth century, was readily accepted even in the absence, to this day, of an entirely satisfactory causal theory of ice ages. We saw the opposite occur with biological evolutionary theory: Lamarck had a plausible causal evolutionary theory (for its time, even though it was ultimately proven false), yet he failed to establish the basic notion that life has evolved. Darwin succeeded, despite the difficulties many of his contemporaries and successors had in fully accepting natural selection as the motor of evolutionary change.

The difference between Darwin and Lamarck, rather, lay mostly in their choice, and handling, of pattern. And so it seems to be with Wegener. True, process seemed to be the essential stumbling block: The earth was too solid to allow plastic deformation, and tidal forces were too weak to produce movement. But as soon as geophysicists encountered *their own* patterns, which *could not be ignored* because those patterns arose from their own geophysical data, they were simply forced to abandon their once monolithic opposition to notions of relative crustal displacement. Yet, such patterns were fundamentally no different than those for *Mesosaurus* and *Glossopteris*.

Wegener's patterns, on the other hand, could safely be ignored or explained away, not just because he was an outsider to the profession, but because he severely mishandled some of his patterns, his "arguments." We have already encountered one glaring example: Wegener's acceptance of shaky longitudinal measurements, which caused him to propose what even at the time seemed ridiculously fast rates of continental displacement. One of Wegener's most frequently mocked patterns resulted from his willingness to consider such fast rates. It involved his examination of terminal moraines, the linear piles of debris plowed by the icy fronts of advancing glaciers, which are left in a heap when the glaciers retreat. Wegener persuaded himself that mapping the terminal moraines of the continental glaciers of the Pleistocene in both Eurasia and North America would make more sense if done on a map with North America much closer to Europe than it now is. But even in the 1920s, geologists knew that the four major glacial advances of the Pleistocene happened relatively recently (now known to be within the last 1.65 million years). Europe and North America have not drifted all that much farther apart in such a relatively brief span of geological time. Justifiably, geologists of the twenties rejected out of hand Wegener's claims of such fast rates and the supposed relevance of the Ice Ages to documenting the reality of continental drift.

Then, too, Wegener's handling of "polar wandering," which was later to prove a decisive point in his favor, was rather dicey. The magnetic poles have long been known to "wander" around the true axis of rotation. However, prevailing opinion, then as now, is that the poles don't actually go anywhere far. They may wander, but they do so within a fairly circumscribed circle. Its outer limits are never very far away from the rotational pole.

Wegener thought otherwise, but his evidence was entirely circumstantial. He looked at patterns of paleoclimate, noting evidence for ancient ice ages in the Upper Paleozoic of South America, southern Africa, and India. The supposition was that each of those land areas was located in the higher southerly latitudes in the Upper Paleozoic. That is, it was positioned where glaciation could take place. Assuming that climatic belts remain pretty much as they are today, such evidence of disjunct climatic zones in the past can make sense only under the assumption that sectors of continental crust have been moving around. There is no other way a place like southern Africa can develop a polar climate. Or so Wegener thought.

So, for that matter, do we think today. But the vast majority of Wegener's contemporaries didn't think so. The Permian glaciation was dismissed as only part of a pattern, the complaint being that Wegener only emphasized the southern hemisphere. He ignored evidence that more-or-less contemporaneous glaciation had occurred in North America and Europe, which were said to have had tropical climes at the time.

Wegener did recognize clear-cut, unambiguous patterns, especially biological and geological historical patterns, but also the haunting matchup of continental margins. But frequently he mixed them up with more ambiguous patterns. And these patterns were all too easily dismissed. For example, the biogeography of living plant and animal groups, especially in the context of his exaggerated claims of rate of continental displacement, was quickly rejected. Then there were his outright absurdities, such as invocation of Pleistocene glacial moraine distributions as evidence of drift. So geographers, biogeographers, and paleobiogeographers persisted in invoking Suessian land bridges to explain faunal and floral similarities: *Mesosaurus* could be found in the Carboniferous rocks of Argentina and South Africa because the two continents were indeed connected, but by an hypothesized land bridge, and *not* because South America and Africa were ever snuggled up together, cheek by jowl, as Wegener claimed. Or so it was thought.[9]

George Gaylord Simpson, one of three people to whom this book is dedicated for his conviction that paleontological pattern has much to tell us about how the evolutionary process works, to his discredit never really did embrace plate tectonics before his death in 1984. Nor did Harold Jeffreys, who had led the geophysical charge against Wegener in the 1920s. Though there is something, I suppose, to be said for sticking to one's guns, and to eschewing bandwagons, I greatly prefer the mental flexibility of scientists like Marshall Kay. About the same age as Simpson, he was able to change nevertheless with the times even though those times show that superficially at least, the paradigm under which he spent most of his working life was fundamentally wrong. Kay simply asked: What do we do now to interpret the facts of geology as we know them under this new paradigm of plate tectonics? That is science at its best.

One Man's Pattern: The Advent of Plate Tectonics

Wars can have curious side effects. Research directed to the detection of enemy aircraft during World War II led to the emergence, in Britain, of radar. In addition, a great deal of effort was given to the problem of detecting submarines. The result was that much was learned of the physics of sound propagation in seawater. These advances were later to have enormous implications for purely exploratory, scientific purposes in the 1950s and 1960s. The fifties and sixties were an age of team science. Large-scale application of new technology, aided especially by the

International Geophysical Year (1957–1959), became a matter of costly exploration of the oceans and atmosphere. Many scientists were involved, sometimes from a single large research institution, but often involving the collaboration of national and international teams of scientists from two or more such institutions.

Among the technological devices developed during this time were sensitive instruments capable of detecting very slight perturbations of magnetization: magnetometers. Rocks with high iron content—especially igneous rocks like basalts, but also red sedimentary rocks (whose color comes from quantities of oxidized iron)—retain the magnetic orientation that was in place at the time they were formed. In the case of basalts, the iron-rich particles line up and are frozen in place as the lava cools. Much the same occurs when sediments rich in iron accumulate.

Two very different sorts of patterns emerged as geophysicists using sensitive magnetometers systematically charted the ocean depths, and as they examined patterns of so-called remanent magnetism in terrestrial rocks. In the latter case, geophysicists such as Cambridge University's S. K. Runcorn and his colleagues discovered in the mid-1950s that the north magnetic pole, as viewed from Europe, had shifted from a position near present-day Hawaii in the Precambrian, to a place north of Japan, before finally coming to its present position in the Tertiary.

Wegener, of course, had invoked polar wandering as evidence of continental drift, only to be derided for it. But in his day no such instrumentation capable of measuring remanent magnetization was in place. In fact, Wegener had inferred that the poles had shifted position strictly through his analysis of patterns of faunal and floral distribution, revealing, he thought, shifting positions of the tropics. So he had had to postulate that grand-scale polar wandering was plausible based on observations of relatively slight short-term changes in mapped geomagnetic pole position. And because no one knew then (or even knows now, precisely) why the earth acts as a dipole magnetic (that is, what sets up the force field), no one was sure that the earth had *always* acted as a simple dipole magnet. It seemed reasonable to most geophysicists in Wegener's day, and it still does, that the measurable movements of the position of the magnetic poles is merely minor fluctuation and not part of some grand-scale pattern of polar wandering.

Then, too, there is the issue of what is really moving: Are the poles *really* moving, or are the continents themselves moving? Both would yield the same result: Poles in the past were not located where they are today. That either the poles or the continents moved had to be the case, unless the mid-1950s data were somehow in error. Many were no doubt surprised, especially those holding to the stabilist position, which by then was so well-ensconced in geophysics. Either answer implied that the solid earth was perhaps not so solid after all.

Once again, it was Runcorn and his colleagues who provided the answer to this conundrum with an elegantly simple, definitive experiment. As shown below, if it had been the poles themselves that had moved, Runcorn's group reasoned, exactly the same path of "wandering" detected in the European paleomagnetic data should show up on other continents. If, instead, it had been the continents doing the moving—vis-à-vis one another as well as the north magnetic/rotational pole—then each continent should display a *different* pattern of polar wandering through geological time.

So Runcorn and his team went to North America. Their results were straightforward but also earth-shattering. They found a curved path of polar wandering, from the Precambrian to the Recent in North America that was offset from, but exactly paralleled, the European track until the two converged near the present-day north pole in the Tertiary.

Assuming there was always one north pole, so that the polar wandering tracks really must have coincided through geological time, Runcorn's team realized right away that the differences between the

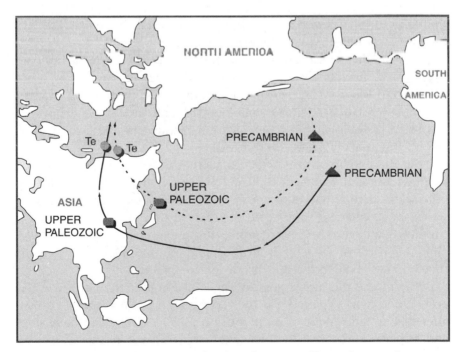

A simplified sketch of Runcorn's diagram depicting offset polar wandering curves as measured in Europe (dotted line) and North America (solid line).

European and North American polar wandering tracks disappear when the two continental masses are reconstructed as lying side by side. In the Devonian, when the two continental landmasses were one, the iron grains lined up and pointed to the same north magnetic pole. Because both Scotland and eastern North America have moved, the orientation of the grains in their Devonian deposits no longer point to the actual, permanently fixed north geomagnetic pole. And because these two land bodies have moved away from each other, the Devonian magnetic results differ, systematically, from each other as well. And, just as modern compass readings in Scotland and New York agree with one another on the location of the north geomagnetic pole, as the continents came closer to their current positions paleomagnetic readings of progressively younger rocks reveal progressively converging positions of the pole.

These results were stunning. But they were soon to be supplemented by an equally remarkable set of oceanic-derived paleomagnetic data. Oceanographers in institutions like Woods Hole Oceanographic Institution, Lamont Geological Observatory, and Scripps Institute of Oceanography had turned to detailed mapping of the world's seafloors. Attention was especially riveted on the 30,000-mile-long rift valley system: the mid-Atlantic ridge, its connections around the world, and its similarity to, and connection with, such terrestrial systems as the East (and now southern) Africa Rift Valley system.

Talk about pattern! A glance at the map of the ocean floors published by Columbia University–Lamont Observatory geologists Bruce Heezen and Marie Tharp reveals a fascinating pattern: linear trackways, looking almost like skidmarks, marking the path of the northeastward migration of India after it broke off from Gondwana and made its way toward eventual collision with the Asian plate.

But the patterns had not yet reached a critical mass for the world's geophysicists and geologists to abandon their traditional, antidrift, stabilist view. It took one more set of observations, historical patterns revealed in geophysical data, to tip the scale and force the acceptance of the new paradigm. And here, unlike the previous evidence, the final, clinching pattern was actually *predicted* and found to be there when the time, money, and effort were expended to look for it.

Part of the routine exploration and mapping of the oceans in the post–World War II period was the recording of slight magnetic anomalies (slight differences in magnetic intensity) detected by both ship- and airplane-borne sensitive magnetometers. One particularly striking pattern of parallel lines of such anomalies was mapped by an expedition from Scripps, off the northwest coast of North America in the late

1950s. On a map of the seafloor, the anomalies looked like zebra stripes—alternating bands of dark and light, depicting a regular pattern of alternation of magnetic signals.

It was geophysicist Fred Vine who turned these anomalies into the final piece of evidence. Geophysicists had just recently realized that the earth's magnetic field, for reasons still not completely understood, goes through cycles of decay and renewal, weakening effectively to zero before slowly building up again. The entire process may take 10,000 years. When the field builds up strength again, there is apparently a 50-50 chance that the polarization of the field will be the opposite of what it was. In other words, the north pole, say, is positive prior to decay, but it can come back with the polarity reversed, as negative. Thus, we have the magnetic equivalent of "flipping" north and south.

In 1963, Vine and his adviser suggested that the zebra-stripe patterns of magnetic "anomalies" actually correspond to alternating bands of the plus and minus charge from the geomagnetic poles, as shown below page. The sequence of polarity reversals had been worked out to some fair detail for the last 3 million years, primarily through use of terrestrial lava flows. The pattern also matched the sequence of magnetic reversals documented in the growing number of deep-sea sediment cores that had been accumulating as yet another aspect of the postwar all-out effort to explore and analyze the world's oceans.

Meanwhile, also in the early 1960s, geologist Harry Hess of Princeton University put forth his theory of *seafloor spreading*. Hess

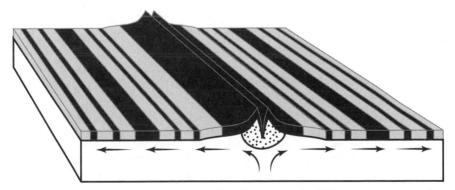

An idealized diagram of the pattern of "zebra stripes" of alternating positively and negatively magnetized strips of oceanic crustal basalt. The stripes form a symmetrical pattern on either side of an oceanic ridge, where new crust forms, cools, sinks and moves laterally.

claimed that the volcanism of the mid-Atlantic ridge represents the up-welling of new oceanic crustal material (sima), which spreads away later-ally, cools, and then, being denser, sinks down to the basic level of the ocean floor. Hess proposed his idea to account for two things: accumulat-ing evidence that the ocean basins are by no means as old as the nuclei of continents, and the fact that, like the African Rift Valley system, the top of the oceanic ridge system is also sharply depressed, as if a central block is faulted downward, which is a sign of tension. To complete the setting of the stage for the last act, University of Toronto geologist J. Tuzo Wilson suggested that the location of major faults, such as the notorious San Andreas of California, and patterns of volcanism outline a series of major blocks of the earth's crust, which he called *plates*.

Hence the prediction: If new crust were being created, flowing to ei-ther side of the midocean ridge system, there should be a *symmetrical* pattern of zebra stripes, now interpreted as alternating strips of plus and minus remanent magnetic polarity. Also, the oceanic crust should be youngest on the midoceanic ridge itself and oldest nearest the conti-nents.

By the mid-1960s, oceanographers were convinced that not only was the pattern of magnetic reversals symmetrical on either side of the ridge but also the patterns themselves match up on either side of oceanic ridges all over the world. Like tree rings, whose variable widths record summers with greater and lesser growth potential, so do strips of polar-ized oceanic crust vary in width. Wider strips simply conform to longer periods of time between reversal episodes, as already abundantly docu-mented in terrestrial lavas and in the sediments of deep-sea cores.

But what about the age of the crust, which should be progressively older as one leaves the ridge and approaches the continental margins? Drilling through the bottom sediments to obtain pieces of oceanic basaltic crust that could be taken back to the lab for absolute dating was the hard way to obtain evidence. With the National Science Foundation's new Joint Oceanographic Institutes Deep Earth Sampling Program (JOIDES), all that was needed was to position the deep-sea drilling ship *Glomar Challenger* on a transect across the South Atlantic, punching holes down to the lowest sedimentary layers resting on the crust. Up came the samples, and the shipboard paleontologist, position-ing a sample under a microscope, could tell in a matter of minutes the age of the sample—based, of course, on the marine microfossils present. It is ironic that, in the end, it took a paleontologist's word to corroborate the new paradigm. After all, those paleontological data supporting con-tinental drift had been so long ignored by the now enthusiastic cadre of geophysicists!

Thus the sudden rush from stability to mobility. This was the beginning, not the finale, of plate tectonics, however. For, while Wilson had already used the term *plates*, the idea of the earth being constructed of a number of rigid plates involving both oceanic and continental crust was not finalized until near the end of the 1960s. Of course, work on the specifics of the theory continues. But much of the basic outlines of plate tectonics—its mechanisms, its consequences for understanding earth history—have at least been sketched in.

Plate tectonics is a marvelous amalgam of structure, function, and history. When, in the early decades of the twentieth century, anthropologists used to debate the merits of a structural versus an evolutionary approach, it seemed to them an either/or proposition. The British anthropologist Radcliffe-Browne took the position that exhaustive description of the components of a society sufficed to capture what that society was. Evolutionists, such as Leslie White, countered that mere description of parts by no means sufficed. It was the evolution of the society—how it got to be the way it appears to be—that really mattered. Probably as a result of such wrangles, evolution in the social sciences has never consistently proven to be the useful approach it most surely could prove to be. In any case, it seems obvious that the internal structure and function of a system must be grasped for its "evolution" to be understood. Also it is apparent that the development ("evolution") of a system is as important to understanding it as the simple description of its components.

I noted at the outset of Chapter 2 that the resonance between historical pattern and geological process inherent in the initial work of James Hutton was quickly lost as geology professionalized. Geologists either worked out the details of history or they looked at ongoing processes such as volcanism, seismicity, and sedimentation. True, Lyell's dictum, "The present is the key to the past," the aphoristic embodiment of "methodological" uniformitarianism, meant that there was still some connection between the two as geologists apply their knowledge of ongoing process to explain events of the past. But the symmetry was broken. Rarely if ever did anyone, confronting an event of the remote geological past, suggest that some additional process, operating at a scale of greater or lesser magnitude than contemplated by those studying processes operating around us at the moment, might have actually operated in the past.

But plate tectonics is a true integrative theory of the earth, one with more than a hint of Huttonism at work. To be sure, a grand initial success of plate tectonics has been a striking reinterpretation of all 4.6 billion years of earth history. Wegener, who for an outsider did so well in marshaling historical pattern to buttress his claim of drifting continents, did not suggest that his theory embraced all of earth history. But the mature version of his theory does achieve precisely that.

But there is more. Geologists are beginning once again to take pattern and data of earth history as not only marking the inexorable accumulation of well-worked-out dynamics of change, such as the 2-centemeter-a-year rate of spreading of oceanic crust, but also providing evidence of remarkable, unexpected events and phenomena. Collisions between the earth and extraterrestrial objects (*bolides*—comets and asteroids predominantly) are a case in point. If, as craters suggest, the earth has been occasionally struck by such objects, it was nonetheless a novel suggestion in extinction theory that comets may well lie at the heart of the mass extinction at the end of the Cretaceous. Such a suggestion sheds light on the role of habitat destruction in triggering extinction episodes. Hence it is helpful in assessing current events as humans transform the world's ecosystems, thereby driving tens of thousands of species to extinction every year.

Still there is nothing that would lead us to anticipate the saturation bombardment of the earth that must have occurred just less than 3 billion years ago. Nothing, that is, save the clustering of the moon's bombardment at the same time. More remarkably, California Institute of Technology geologist Joseph Kirschvink and colleagues,[10] at this writing, is suggesting two very remarkable, wholly unanticipated events in early geological history: *snowball earth*, where massive glaciers invaded even the tropics twice, at 2.2 billion and 800 to 600 million years ago. Both events are correlated with major events in the evolutionary history of life: The former was followed by the appearance of the first complexly celled organisms, the eukaryotes, consisting of fungi, protoctists, and algae; the second was followed by, and may well have triggered, the evolutionary diversification of the major animal phyla.

More remarkable still is Kirschvink's claim that the Cambrian "explosion," the near-simultaneous appearance of a diverse animal fauna marking the advent of a rich fossil record, followed an amazing geophysical event: The entire lithosphere (that is, the earth's crust and upper mantle) rotated as a shell almost 90 degrees, "disrupting climatic zones and stirring the evolutionary pot."

Fifty years ago, when such radical ideas were mooted, they were dismissed as crackpot. The fate of Kirschvink's claims is not yet settled but, based on historical evidence and the 20-year-old realization that the earth is internally plastic, Kirschvink's suggestions are taken seriously indeed.

Thus the plate tectonics revolution has given us an integrated theory of the earth. It has also brought historical and dynamical geology back more nearly to their original, Huttonian formulations as mutually illuminating, co-equal partners. Patterns of earth history have much to tell us about the forces that shape it.

Emerging, too, is the strong realization that the evolution of life is intimately linked to the physical history of the planet. It becomes evident in the following chapter that not even the modern history of evolutionary biology seriously examines, let alone provides the answer to, the question: What is the relation between energy sources, between matter-in-motion and the biological evolutionary process? The final chapter looks at the relation of earthly dynamics and evolution as part of the answer.

We have seen in this chapter that Hutton is proven right once again in terms of geology's own issues of "matter-in-motion." For important as solar energy is to the generation of motion in water and atmosphere, and as important as sheer gravity also is in shaping the earth's exterior, it is heat flow from the earth's interior (produced by atomic decay and not, as earlier supposed, as a residuum of a once fiery, molten state) that drives the plates around. Interestingly, no geologist claims to understand perfectly how this process works, though some form of convection must be operating. Compare this state of affairs with evolutionary theory where no formal links with energy flow have ever really been forged. Nevertheless, most evolutionary biologists will say, without hesitation, that the central mechanism of evolution, natural selection, is in itself completely understood.

CHAPTER 5

Of Genes and Species:
Modern Evolutionary Theory

Darwin had triumphed completely in his central objective: to convince the thinking world that life has had a very long history, a history of evolution from a single common ancestor. He hammered home the importance of continuity in pattern, and focused purely on genealogical systems. For that is what evolution produces: skeins of ancestral and descendant species.

If Darwin's notion of *how* evolution occurs, natural selection, provoked more criticism than the overarching notion that evolution has occurred, nonetheless it too became generally accepted as *the* mechanism of evolution. But natural selection was then a very different sort of scientific generalization, or "law," than most mid nineteenth century scientists were accustomed to. For that matter, it remains today very different than what physicists, chemists, and geologists are accustomed to. Theirs is a world of matter-in-motion, where the interactions of physical entities are customarily summarized by equations, not always entirely successfully or even neatly.

Darwin's law of natural selection was unique: a generalization about the ineluctable bias in transmission of what we now call *genetic information*. Remarkably, this statement was formulated some forty years prior to the advent of a coherent science of inheritance—genetics. And information theory wasn't to blossom for at least a hundred years. The principle, as we have also seen, still stands, after thorough evaluation of theory, in the lab and in the field.

But Darwin's focus on pure genealogical systems, though thoroughly understandable, kept evolutionary biology at some remove from the world of matter-energy transfer processes. Only a few hints of how those connections might be drawn are latent in his work. These include his distinction between true "natural" selection and sexual selection, and his emphasis on physical factors controlling the size of populations—thus generating the pattern of differential reproductive success that lies at the heart of natural selection. In short, most organismic adaptations are *economic* in nature, meaning they have to do with obtaining energy or avoiding becoming someone else's meal.

The early decades of the twentieth century saw Darwinism in retreat. The early triumphs of genetics were taken either as directly refuting Darwin's postulates (for example, on the continuity of evolutionary phenomena), or at the least as rendering natural selection superfluous in the search for evolutionary mechanisms. The importance of genetics in understanding the evolutionary process is, of course, profound. But the hyper-reductionism of theorists such as Richard Dawkins (see the discussion at the end of Chapter 3) has exacerbated this lack of conceptual linkage between the inorganic, physical realm and biological systems. The world of matter-in-motion and the intergenerational transmission of genetic information remain separate concepts. Twentieth-century biological history can by no means be read as a simple progressive story of building bridges between these two realms. Though some progress has been made, we still have a long way to go.

Darwinian Eclipse: The Advent and Early Triumphs of Genetics

Biology entered the age of modern science when electricity came to the laboratory around the turn of the century. The old stigma of a field-oriented, observational natural history quickly was replaced by a renewed optimism and enthusiasm for the mysteries of physiology. In particular, an emphasis was placed on inheritance, which was always intuitively acknowledged to be processes engaged in by unseen microscopic structures lying within the cell.[1]

In the years immediately following the 1859 publication of the *Origin*, embryologists, especially the German Ernst Haeckel (1834–1919), rushed in, reinterpreting diagrams of similarity among organisms (such as the old *scala naturae*) as evolutionary trees. Haeckel imagined that in a sense a developing embryo retraces the evolution of its lineage. Similarly, when traced back, the early stages of ontogeny (the development of an embryo) of related organisms seem to converge: Early embryonic stages of mice and men are far more similar than are the fully formed organisms at birth. From this came the famous cry "ontogeny recapitulates phylogeny." To the extent that this statement is true, it remains a useful source of both data and insight in the ongoing task of elucidating phylogenetic history.

But embryological pattern, no matter how elegantly its details are revealed, is frustratingly mute on the essential mysteries of how eggs and sperm from sexual organisms that are *conspecific*, or of the same species, continually develop so faithfully into a full-fledged new member of that very same species. How do organisms develop from fertilized eggs? And just what are the factors that cause offspring to resemble their parents? Darwin's formulation of natural selection did not require anything more than the observation that organisms do indeed resemble their parents. Biologists in the latter half of the nineteenth century understandably felt that, were the secrets of inheritance to be revealed, evolution itself would come to be seen in an entirely new light. After all, if evolution is simply intergenerational change in heritable variation, understanding how heredity works in the first place, including understanding how *new* variation arises, might be all the knowledge we would need to understand how evolution works. Or so it seemed to many geneticists at the turn of the century.

While a true science of heredity continued to elude biologists in the latter half of the nineteenth century, progress had nevertheless been made. Gregor Mendel (1822–1884), the Austrian monk who experimented with patterns of heredity in pea plants in the mid-1800s, leaps to mind first. Yet the significance of his work lay unrecognized until its rediscovery at the turn of the century. Prior to that, the most important step forward in what was soon to become genetics, particularly from an evolutionary point of view, was a realization by the German biologist Auguste Weismann (1834–1914) in the 1880s. He determined that inheritance was effected solely by the sex cells, for example, by eggs and sperm in sexually reproducing animals. Weismann recognized an important and unequal cellular dichotomy: *Germ-line* (or *sex*) cells create both the sex cells and the *somatic* (or *body*) cells of offspring individuals. The implication is enormous: *What happens to the cells of an individual's body, its somatic cells, during its lifetime cannot affect its sex cells, which are present from an early stage in development.* In a single blow, Weismann removed the possibility of the inheritance of acquired characteristics. That notion had been the linchpin of Lamarck's evolutionary mechanics, and was adopted even by Darwin in the sixth edition of the *Origin* as he tried to placate his critics.

Weismann's dictum of the separateness of the soma and the germ-line—and that the sex-line determines the soma of offspring but the soma can have no effect on the germ-line—has worn rather well. It lives on as the "central dogma" of molecular biology: Ribonucleic acids RNA and DNA, the molecules of heredity, act as templates for the production of proteins, but proteins do not in turn dictate the structure of ribonucleic acids. Like all such generalizations or "laws," both Weismann's original dictum and the central dogma have interesting, although hardly fatal, exceptions.[2] The inheritance of acquired characters remains a functionally dead issue in evolutionary biology.

Weismann's distinction between germ-line and soma also has links with the Darwinian past. Natural selection involves evolutionary changes in the soma, while sexual selection affects reproductive structures, including the gonads and those aspects of the soma with obvious reproductive functions—Darwin's "secondary sexual characteristics." As we shall see in Chapter 6, Weismann's distinction has even further implications for the formal analysis of structure and functions of biological systems generally. The somatic aspects of organisms are what link them to the physical environment, the world of matter-energy transfer, while it is the germ-line that records what works better than what. Keeping score of information *is* evolution.

But it was the near-simultaneous rediscovery of Mendel's work that set off the cascade of research at the turn of the century. Overnight the science of genetics was born. Mendel's "laws"—generalizations based on repeated patterns of inheritance—were ideas whose time had long since come.[3] This much is obvious from the famous circumstance in which they were rediscovered. It was not once, nor even twice, but three separate times: by the botanists Hugo De Vries, Carl Correns, and Erich Tschermak, each of whom published his findings in the spring of 1900.

Mendel established that the elements responsible for heredity are particulate and come in alternate forms. In the first decade of the twentieth century, those particulate hereditary factors were identified and named *genes*, each with its own position, or *locus*, on a chromosome within the cellular nucleus.[4] This discovery led to a particularly—and literally—graphic result: microscopic images of stained giant chromosomes from the salivary glands of fruit flies revealing consistent banding patterns. Such patterns could hardly be interpreted as anything but the existence of genes occupying regular positions along the chromosomes.

But the very particularity of genes seemed at odds with Darwin's insistence that variation in nature is smoothly continuous. Mendel worked with simple systems where, for example, the skin of a pea would either be "smooth" or "wrinkled." The alternate forms of the genes (*alleles* in modern parlance) governing smooth and wrinkled exhibit a pattern of *dominance:* When both alleles for smooth are present (*homozygous*), the skin is smooth; when both for wrinkled are present, the skin is wrinkled. But when a pea plant has one of each allele (*heterozygous*), the skin is wrinkled, as that allele dominates the smooth one. In the next generation, the alleles are sorted out, and the heterozygous plant can be a parent to homozygous dominant or recessives, depending on how its two different allelic forms match up with the alleles from another plant.

In such simple systems, then, there are sharp discontinuities, both at the genetic level and when they are expressed as phenotypic features. So how can natural selection gradually modify the genotypes and phenotypes of such systems?

Then there were the issues surrounding the origin of new, heritable variation—new forms of alleles that quickly came to be known as *mutations*.[5] Darwin had already written of the spontaneous appearance of heritable novelties, but the mutations that caught the attention of the earliest geneticists were overwhelmingly negative, or *deleterious*. A function primarily of as-yet-unperfected laboratory techniques, those mutations easiest to spot were ones that were downright lethal, caused visible monstrosities (a double set of wings in fruit flies, for example), or otherwise produced obviously dysfunctional individuals. Hardly the stuff of new novelties leading to progressive evolutionary change! Only later, as the early decades of the twentieth century wore on, were mutations with smaller effects detected. These mutations were only slightly deleterious, had no noticeable effect one way or the other ("neutral"), or actually seemed to produce more viable offspring.

It was one of Mendel's rediscoverers, the Dutch botanist Hugo De Vries, working on the evening primrose, who made the grandest claims concerning mutations. (The evening primrose is one of the few plants that has spread from North America to Europe as a weed, which is the reverse of the more common, post-European colonization pattern.) The "mutations" De Vries encountered in his observations on evening primroses involved fairly gross changes in the morphologies of the flowers, changes on the order of magnitude of the differences between closely related species. Perhaps, De Vries suggested, new species themselves arise overnight, by simple acts of mutation.

Gone, in one fell swoop, were Darwin's hundreds of thousands, if not millions of years, of slow, steady evolutionary change in the emergence of new species. Eventually De Vries's primrose mutations turned out not to be simple "point" mutations at a single genetic locus but rather substantial chromosomal rearrangements. Nevertheless, the idea that large-scale overnight mutational change can be important in significant, and obviously rapid, evolutionary events has checkered twentieth-century evolutionary biology. These *macromutations* have never been popular with the majority of geneticists—who understandably point to a lack of confirmatory experimental evidence.[6]

In short, the early results in genetics, coming so fast, and being so hard to digest, really did seem to throw a lot of doubt on the soundness of Darwin's dicta of continuity and slow, steady change mediated by natural selection. By the late teens and early twenties, biology had become a Tower of Babel insofar as discussions of evolution were concerned: Every contributor seemed to favor one different version or another. Things were so bad that Henry Fairfield Osborn, simultaneously president and director of the American Museum of Natural History and chairman of the Zoology Department at Columbia University, saw fit to invent his own theory of evolutionary genetics. Though a professed ardent admirer of Darwin, Osborn imagined progressive evolution to be the result of inherent tendencies toward improvement latent in the genome itself. He called this notion *aristogenesis*. It is no coincidence that Osborn was a very wealthy man.[7]

But organisms seem to fit their environments exceedingly well. And Darwin's notion of evolution through natural selection accounted for this fact better than any other theory. The Darwinian vision never disappeared so much as a weedlot of extraneous, supposedly competing ideas grew up around it, for a time choking out the simple clarity of Darwin's original vision. Small wonder that, sooner rather than later, a small army of genetically sophisticated biologists appeared to weed out the plot. They effected a true and lasting rapprochement between genetics and Darwinian evolutionary theory.

Weeding the Lot: Fisher, Haldane, and Wright

Pattern, not surprisingly, prevailed. Perhaps there really is progress in the growth of knowledge, and maybe hard-won, accurate descriptions of the way things are will, in the end, win out! It soon became clear that nothing encountered in the genetics laboratories of the first two decades of the twentieth century could plausibly be interpreted as an alternative to the Darwin-Wallace concept of adaptation through natural selection. And organismic design, remember, is the original problem that motivated Darwin in the first place.

Three names stand out in the move to seek common ground, to effect a true rapprochement between the rapidly accumulated solid body of knowledge in genetics and the older, but obviously still valid, visualization of evolutionary dynamics achieved in the previous century. Foremost among them, perhaps, was the Englishman Sir Ronald Fisher (1890–1962), a mathematician by training and bent, but a man who, perhaps more than any other is responsible for formulating the hard inner core of evolutionary theory, what is often called the *neo-Darwinian paradigm*. The other two were the British polymath J. B. S. Haldane (1892–1964; mentor to John Maynard Smith, dean of what I call the ultra-Darwinian school of evolutionary theory) and the American Sewall Wright, of whom more later. In a series of papers beginning right after World War I,[8] these three mathematically inclined scientists managed to cut through the confusion and many conflicts and contradictions between genetics and by-then-traditional Darwinian evolutionary theory.

What these three pioneers saw went straight back to Gregor Mendel: Alleles come in *frequencies*. Mendel, for example, found ratios of 3:1 in wrinkled versus smooth peas. He realized that this meant a ratio of 1:2:1 of homozygous dominant (1), heterozygous (2), and homozygous recessive (1). As far back as 1908 two mathematically inclined geneticists, G. H. Hardy and W. Weinberg, had established that such allelic frequencies are destined to remain constant (the *Hardy-Weinberg equilibrium*) unless and until disrupted by immigration, mutation, natural selection, or non-random-mating behavior.[9]

Fisher, Haldane, and Wright were the founders of *population genetics*, the mathematical analysis of the fates of gene frequencies. Population geneticists explore what happens to gene frequencies under certain conditions as pure mathematical theory. Alternatively, they analyze changes in frequencies in experimental populations in the laboratory—or in populations in the wild. Fisher was also a key figure in countering what seemed to be one of the strongest objections by geneticists to Darwin's patterns of continuous variation (on which selection could work incrementally). He was able to convince the world that continuous variation resulted from the additive effects of many different genes working in concert to shape a particular feature such as an organism's height.

It was Ronald Fisher above all who re-established the primacy of Darwinian natural selection as the central and most important agent of evolutionary change by far. To Fisher, and to the vast majority of population geneticists who have followed him, natural selection is, at once, both necessary and sufficient to account for evolutionary change. Thus the core neo-Darwinian paradigm: Natural selection effects evolutionary change by working, within populations of organisms of the same species, on a groundmass of genetic variation, the ultimate origin of which is mutation. Mutations arise. If they are beneficial, they are immediately selected. If neutral, they are tolerated and accumulate: Selection is "blind" to neutral mutations. If they are negative, they are selected against. But the overwhelming rule is that mutations are not themselves adaptive responses to the agent of change, even if they are environmentally induced, for example through exposure to radiation. They are simply adventitious changes in genetic coding—changes whose fate sooner or later will be adjudicated by natural selection.

I have already remarked that the essence of Fisher's wisdom survives, albeit in advanced and much more sophisticated form, in the works of such modern authors as George Williams, John Maynard Smith, and perhaps quintessentially Richard Dawkins. Their contributions to evolutionary biology have been impressive.[10]

On the other hand, it has always been clear that population genetics alone will hardly suffice to produce a complete evolutionary theory—one that can tell us why evolutionary change happens when it does and why, for the most part, stability is the rule. It was Sewall Wright, soon augmented by three more luminary biologists, who tackled these issues. He began developing the pathways that would lead in the direction of a more complete and universal evolutionary theory. Such a theory begins to encompass the hierarchical structure of the biotic world. And, above all else, it unambiguously links the realm of matter-energy transfer with the evolution of life. Though the complete theory has yet to be written, we are getting closer, thanks in no small measure to Wright, and to the second triad of Theodosius Dobzhansky, Ernst Mayr, and George Gaylord Simpson.

Wright, Dobzhansky, and Mayr: Emergence of the Biological Species Concept

Over and over, Darwin emphasized patterns of continuity in the living world. Continuity in morphology and its distribution over the earth and through time were precisely the antidote required to combat the rigid separateness of species that had been envisioned by the biblically minded, creationist-inspired early naturalists. Yet, at least some biologists coming hot on Darwin's heels[11] realized that, in pushing the importance of connectivity and continuity, he had in some sense overstated his case. Contemporary species for the most part *do* seem distinct; and closely related species, for the most part, do not show patterns of intergradation. But discontinuity as a legitimate biological pattern that had to be taken into account by evolutionary theorists remained on the back burner until the 1930s.

Theodosius Dobzhansky (1900–1975) changed all that. Dobzhansky came to New York from the Soviet Union in 1927 to work in Thomas Hunt Morgan's by-then-famous genetics lab in Schermerhorn Hall on Columbia University campus. Dobzhansky had been trained in evolutionary biology primarily as a systematist, with an expertise in ladybird beetles (Coccinellidae). His uniquely creative approach to evolutionary biology sprang from the tensions that arose between his experiences as a systematist and field biologist and his adopted discipline of genetics.

Tensions sometimes allow bridges to be built, and new insights are often concentrated along the fracture lines between disciplines. Dobzhansky always insisted that whatever was discovered in laboratory populations (of fruit flies or any other experimental organisms) *must* be confirmed in the field. Only then could results be trumpeted as providing additional insight into the inner workings of the evolutionary process.

Dobzhansky's first major contribution came when he was invited to inaugurate the newly revivified Jessup Lecture Series. Mysteriously moribund since 1910, this series was once a fruitful source of titles for the Biological Series of Columbia University Press.[12] Dobzhansky spoke in the series late in 1936, when he was a junior faculty member at the California Institute of Technology. Remarkably, he was able to produce a full-length manuscript which Columbia University Press (in pre-electronic days!) published as the monumental *Genetics and the Origin of Species* in time to be used in classes in the fall semester of 1937!

From the very outset of this book, Dobzhansky struck at the heart of the matter: Natural selection, he said, works on, and in a sense promotes a continuous spectrum of variation. We see this variation within populations and species, abundantly confirming the Darwinian vision. But something further must be afoot, for there is the until-then-underacknowledged *discontinuity* between species that must be taken into account. Darwin had passed over discontinuity lightly. He attributed the anatomical "distance" between species sheerly to the extinction of past, intervening variation. Not so, thought Dobzhansky. Instead, there must be something inherent in the evolutionary process itself that manufactures discontinuity.

Dobzhansky's introductory musings on the discontinuity between species led to a further insight on the hierarchical organization of biological systems. Alluding to the recent "synthesis" between Mendelian genetics and the Darwinian vision of evolution through natural selection, Dobzhansky had the insight to proclaim that genetic information is indeed discontinuous at the level of the individual. Genes themselves, with their allelic forms and the occurrence of mutations, are particulate and discrete. But at the level of the *population*—a collectivity of males and females of the same sexually reproducing species—that particulateness disappears. One talks, instead, of gene *frequencies*. To Dobzhansky, the apparent incongruities between genetics and Darwinian continuity was resolved simply by noting that the two were addressed to different levels. Physiological genetics addresses the level of the individual, and to that individual's allocation of genes in different cells, while population genetics addresses groups of reproducing organisms.

To this neat dichotomy, Dobzhansky then added the next-higher level: species. Species, like mutations, appear to be discrete. Darwin had shied away from this theme because, in his time, the higher priority had been to establish continuity in the living world. He simply had to establish the basic principle that life had evolved. But if the discreteness of genes, their alleles, and mutations had been reconciled with Darwinian natural selection, could not this higher-level discreteness among species now also be accommodated in an expanded Darwinian evolutionary vision?

Dobzhansky certainly thought so, and he set out to accomplish the task. Thus he first spelled out a series of *isolating mechanisms*, ways and means by which a species may be pulled apart to form two or more descendant, daughter species. Foremost among these mechanisms was geographic isolation. Dobzhansky recognized the evident truth of what some post-Darwinian biologists had earlier realized: Geographical variation, especially when disrupted by actual physical discontinuities, or isolation, between closely related populations, was a powerful stimulus to evolution.[13]

Hot on Dobzhansky's heels was Ernst Mayr, who had a set of better examples drawn from extensive field and laboratory work on birds. Mayr, like Dobzhansky, was a European émigré, coming to the United States in the 1920s, when he became a junior curator in ornithology at the American Museum of Natural History. His mentor in Germany had been E. Stresemann, who, along with another ornithologist, Otto Kleinschmidt, had been stressing the importance of geographic differentiation and isolation in the evolutionary process. Drawing on his continental European roots, Mayr fleshed out the theory of *allopatric* (that is, geographic) *speciation*. This theory generalized that new species arise from old, as the overwhelming rule, only after an ancestral species becomes fragmented; if those fragments diverge sufficiently in isolation, such that they can no longer successively breed should they ever again come in contact, two or more full-fledged species would then exist instead of the original single one. In his *Systematics and the Origin of Species* (1942), Mayr laconically pointed out that Darwin, in fact, never even addressed the "origin of species," despite the title of his book. Instead he was concerned with denying the discreteness—thus, to Mayr, the reality—of species.[14]

But if species are real, what manner of entities are they? Here Dobzhansky and Mayr created a profound change in biological thinking. Species were originally the lowest division on the Linnaean totem pole. Despite the addition of subspecies, races, and other subdivisions, they have basically retained that status in the minds of most biologists. So, while it has sometimes been heard that species are "whatever a competent systematist thinks they are," the prevailing presumption ever since

Linnaeus's time saw them simply as groups of similar organisms—organisms more similar to one another than to the members of any other groups of internally similar organisms constitute a species.[15]

That genetic inheritance underlies the pattern of internal similarity defining any particular species was always tacitly recognized. But it was not until Dobzhansky saw that the time was ripe for integrating the evident discreteness of species with the new genetics that a conceptual switch was made. Species were reconceptualized as reproductive communities, a notion pioneered by Dobzhansky but clarified greatly by Mayr. The "short" definition of species by Mayr is justly famous: "Species are groups of actually or potentially interbreeding natural populations, which are reproductively isolated from other such groups."

With Dobzhansky and Mayr's work, then, species became reproductive communities. Its members, not surprisingly, tend to resemble one another more than they do members of other reproductive communities. The logic was tight indeed, as species now became distinctly *genetic* entities. As such, they form an important rung in the hierarchy of evolutionary biological entities: genes/organisms/populations—and now species. Genes are parts of organisms, which are in turn parts of populations. And populations are parts of species. Though Dobzhansky noted this hierarchical structure only in passing at the very outset of *Genetics and the Origin of Species*, he really is the architect of what has come to be termed the *genealogical hierarchy*.[16]

It is ironic that, among the many explicit debts that Dobzhansky owed to the trailblazing Sewall Wright, is his conception of the internal structure of species. Wright was a mathematically inclined theorist[17] and a numerical analyst of experimental data. He was definitely not the sort of field man Dobzhansky was. Yet the picture Dobzhansky paints of the internal anatomy of species comes straight from Wright, who saw species as broken up into quasi-autonomous local populations. Each such population would have somewhat different evolutionary pressures on it, the sorts of factors earlier identified as necessary to disturb Hardy-Weinberg equilibrium, to change gene frequencies and "cause" evolution. Thus each local population would start out with a different subset, a different *sampling*, of the spectrum of genetic variation found within the species as a whole; it would have, on average, a different history of mutations; and it would have a different *selectional history* from all other populations within the species as well.

The latter point is particularly important. Seeing that each local population (which Wright dubbed a *deme*) would be subjected to slightly different regimes of natural selection, Wright understood perfectly well that conditions differ through the range of any given species. Put in more modern terms, local populations of species are integrated into local ecosystems. And physical environmental factors, as well as the identities of the populations of other organisms forming a part of each local ecosystem, will be different from place to place within the total range of any given species.

Here is a vital connection linking genetic systems (local breeding populations within species) with matter-energy transfer systems, specifically, local ecosystems. The point was not strongly grasped or developed by Wright, Dobzhansky, or any of their contemporaries. Nonetheless, it was there, and we will come back to it in the following chapter.

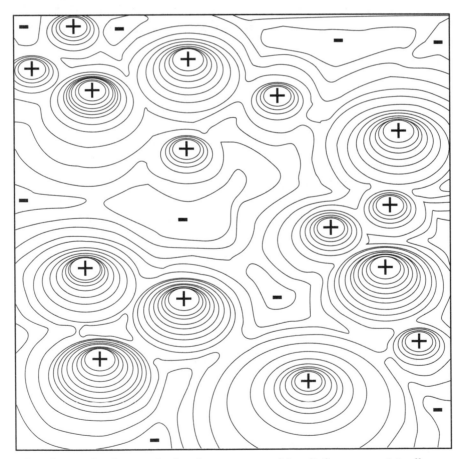

The adaptive landscape, depicted as a topographic relief map, as originally conceived by Sewall Wright and utilized by Theodosius Dobzhansky. "+" denotes hills; "−" denotes valleys.

To Wright, evolution is all about the maximization of what he called the more *harmonious* gene combinations within a species. He came to this notion after seeing the enormous amount of genetic variation that exists within every species and having the certainty that some combinations of these genes are better, more harmonious, than others (that is, they make better, healthier organisms). He developed a simple pictorial metaphor to dramatize the process: a three-dimensional topographic map where the hills are the better gene combinations, the valleys the worse ones. This was Wright's famous *adaptive landscape* as shown in the figure on the facing page. Though he saw natural selection as important to the process of maximizing the better combinations, he also showed mathematically that alleles could go to "fixation" (cannot be eliminated) by sheer chance alone.[18]

Wright quickly extended his metaphor. At first his map of peaks and valleys applied to just the relative "goodness" of gene combinations. But in the very same paper in which he first presented his adaptive landscape,[19] Wright also spoke of entire *species* as occupying "adaptive peaks." Yet it was Dobzhansky who was to extend Wright's peaks and valleys into a universal metaphor for the entire evolutionary process. Though he scrupulously reported Wright's original concept in all three editions of *Genetics and the Origin of Species* (1937, 1941, 1951), Dobzhansky also quickly came to see species as themselves occupying adaptive peaks. But it was issues of large-scale evolutionary history, so-called macroevolution, that prompted Dobzhansky's greatest stretch of Wright's metaphor. In a passage often quoted in recent years,[20] Dobzhansky made further reference to the hierarchical structure of biological systems by extending the imagery to embrace genera, families, classes—in short, all the larger-scale elements of the Linnaean hierarchy. What he said, in effect, was that just as species occupy adaptive peaks, groups of related species form series of hills in the adaptive landscape; viewed from even farther away, those series of hills are seen to be parts of entire mountain chains. For the cat family, for example, Dobzhansky proposed that the niches of all cat species are so similar that they form a discrete cluster. The family Felidae would constitute a range of adjacent peaks, suggesting that the cat family as a whole plays a direct economic role in nature.

Though the two paragraphs of this fleeting passage appear very early in Dobzhansky's text, and though they are not referred to again, they are profoundly important. They indicate Dobzhansky's attitude that *species* occupy *ecological niches*. Moreover, higher taxa also have the equivalents of "meganiches." And, though he has since retracted it, Ernst Mayr himself added an economic component to his biological species concept when he declared that species have ecological niches.

The suggestion that species and higher taxa play concerted roles within specifiable economic systems—that is, have niches—has not fared well. (We will discuss reasons for this in the following chapter.) Nonetheless the attempt to link evolutionary process with the physical world is noteworthy. Dobzhansky, and later Mayr, were responding to a need to fill the evident void: the gap between the evolutionary process and the physical environment. In so doing, they sowed the seeds for more accurate discourse along these lines.

Dobzhansky, then, took patterns of discontinuity between species as "real." That is, he saw them as a direct outcome of the evolutionary process rather than an artifact of differential extinction where the intermediates simply drop out. He used his extended version of the adaptive landscape to highlight the problem: Natural selection would be expected to optimize a species, focusing its adaptations and thus its genes narrowly at the very tip of the adaptive peak. Too much genetic variation interferes with that focus, that adaptive concentration. So it stands to reason that selection will act to fragment a species into two discrete entities, each of which can focus on its slightly different adaptive peak.[21] Dobzhansky attempted to characterize just how evolution acts to isolate species, developing a model of reinforcement, where selection acts against hybrids to "reinforce" the partially emplaced "isolating mechanisms" between hybridizing populations of partially divergent species.

South African geneticist Hugh E. H. Paterson has taken a subtly, but profoundly, alternate view of both the very nature of species and the manner in which ancestral species split and go their separate ways.[22] To Paterson, every species has a unique set of reproductive adaptations—structural, behavioral, and chemical devices that allow males and females to mate successfully. Paterson called this the *specific mate recognition system* (SMRS). Like the biological species concept, Paterson's recognition concept sees species as reproductive communities. But a stronger point is added: that species define themselves, in essence, with reference to a set of reproductive adaptations its members alone share.

Under this rubric, speciation is a pure accident. In isolation, selection (sexual selection!) acts to keep mating successful. But the SMRSs of the isolated populations may easily diverge, whether in response to life in different environments or by pure coincidence. Selection will never act to drive two populations apart. Instead, it will act merely to keep reproduction going within each one of them. Paterson's work suggests that the reproductive and economic adaptations of organisms, in principle at least, change independently in the evolutionary process—testing apart, rather than incorporating, the world of genes and the realm of matter-in-motion.

Shades of James Hutton: George Gaylord Simpson and Quantum Evolution

George Simpson never liked the biological species concept. When asked on a questionnaire devised by Ernst Mayr in preparation for a conference on the origin of the "modern synthesis," "When [did you] first become aware of the importance of the biological species concept?" Simpson snapped back: "I never did."[23] Simpson thought the species envisioned by Dobzhansky and Mayr to be flatly ahistorical. As a paleontologist, Simpson saw species as lineages of sexually reproducing organisms. And as a good Darwinian, he saw such lineages, forever changing as time goes by. True, the lineages themselves are discrete, as only certain plants and microbes exhibit extensive hybridization. Such lineages have their own "separate evolutionary roles and tendencies," as Simpson put it in his own definition of species. But the lineages invariably and inevitably change through time. For that matter, there are passages in Dobzhansky's and Mayr's writing which show that they too, when they thought about it, imagined newly discrete species to gradually transform through time.

What really irked Simpson, then, was not so much the rhetoric on the origin of reproductive disjunction but rather what he took to be the sheer triviality of the entire issue of species. The differences between closely related species are typically minor: Coyotes and timber wolves are fairly similar. To Simpson, all the energy expended on developing a biological species concept was never really worth it, simply because speciation itself entails only minor evolutionary changes. Simpson thought of speciation simply as a process of subdivision of the adaptive "zone" already occupied by the ancestral species—hardly the stuff of evolutionary novelty and innovation.

What grabbed Simpson's attention, instead, were the large-scale changes in evolution: the origin of mammals from an ancestral reptilian stock, the origin of birds, or the diversification of the major subdivisions (the *orders*) of placental mammals. Steeped in Darwinian gradualist tradition, Simpson maintained that countless cases of evolution at and around the species level documented in the fossil record upheld the Darwinian view that change comes slowly, progressively, and gradually—and doesn't get anywhere far. New genera might arise in this slow, steady fashion. But, Simpson claimed in his highly original *Tempo and Mode in Evolution* (1944), such can hardly be the case for families, orders, and all the taxa of higher categorical rank in the Linnaean hierarchy. Some additional factors must be at work that prompt lineages to leave one adaptive zone for another. Such, it seemed to Simpson, was the stuff of true evolutionary creativity.

It was repeated patterns in paleontological data that led Simpson down this particular analytical path. Simpson's goal, beautifully articulated in the preface to his 1944 volume, was to effect a sorely needed rapprochement between his discipline of paleontology and the newly emergent, genetically based Fisherian synthesis. After all, there has been but one evolutionary history of life. And it seemed quite clear to Simpson that genetics had by then, the late 1930s,[24] become a firmly established science of heredity. So too it seemed that the reconciliation between genetics and the original Darwinian vision of evolution through natural selection was itself thoroughly valid. If all these things were true, Simpson reasoned, then the data of paleontology, the paleontological evolutionary patterns, must be consistent with those findings.

But they weren't, at least not obviously, to Simpson or to anyone else who had tried to honestly reconcile the Darwinian vision with the sorts of patterns typically seen in the histories of lineages in the fossil record. I am convinced that Simpson, had he looked more carefully, would never have asserted that species and genera evolve in slow, steady transformations within lineages. He basically went along with prevailing thinking, that such *must* be the case, presumably to show that he was a Darwinian in good standing. But he was after bigger game. Here he really did take a long, hard look at patterns in his and other paleontologists' data.

In example after example, Simpson saw that new groups seemed to appear suddenly in the fossil record. New higher taxa such as whales (mammalian order Cetacea), bats (order Chiroptera), or even the lineage of grass-grazing horses that evolved from leaf-browsing ancestors all made sudden appearances. Seldom was there a long series of intermediate forms that could be traced back through the tens of millions of years that such large-scale evolution would seem to call for.

Moreover, Simpson saw that these new groups first appear pretty much in recognizable form. In modern terms, the defining characteristics, the *synapopmorphies*, that mark a lineage as distinct and evolutionarily homogeneous (*monophyletic*) are in place at the very outset of a group's evolutionary history. Eocene whales, for example, were distinctly whalelike. As one might expect, they were primitive in certain ways as whales; for example, they bore serrated teeth and still retained a pair of pelvic flippers. But those earliest whales were by no means half-way between a four-legged terrestrial mammalian ancestor and a modern sperm whale. They were much more like the latter than the former. Bats offer an even more dramatic example. The earliest ones known, also from the Eocene Epoch, have not only wings but also the distinctive inner-ear apparatus to show that echolocation had already evolved!

And here is the kicker. The earliest whales Simpson knew about are some 55 million years old. If one could devise some sort of measure of rate of evolutionary change,[25] the rate of change within whales over the past 55 million years would seem to be slow to moderate. *If that rate were then extrapolated back to encompass the far greater anatomical changes between the earliest whales and their wholly terrestrial, four-legged mammalian ancestors, we would have to place the beginnings of whale evolution hundreds of millions of years back in geological time!* And that is a patent absurdity, as placental mammals of any kind had appeared at most only a few tens of millions of years prior to the advent of the earliest whales.

Simpson was, in a sense, skating on thin ice. For other paleontologists had also recognized this general pattern. Foremost among them was Otto Schindewolf, a German invertebrate paleontologist. Schindewolf imagined such evolutionary transitions to be as abrupt as the fossil record seemed to be indicating. *Typostrophism* was his term for it, literally a leap between one "type" and another. Such "saltational" ideas are reminiscent of another German biologist, Richard Goldschmidt, the geneticist remembered mostly for his advocacy of the sudden appearance of "hopeful monsters" through "macromutations."

Simpson had to distance himself from such renegade thinking. All the while, though, he admitted, albeit tacitly, that Schindewolf had a point about the pattern of relatively abrupt origin of higher taxa. He did so by developing a theory, *Quantum Evolution*, which could explain the absence of intermediate forms required by any version of a Darwinian-based explanation of evolutionary patterns.

The core of Simpson's idea is that, at certain times and in certain circumstances, lineages evolve in rapid spurts. These bursts occur in such relatively small populations, happen so rapidly, and typically involve such a drastic transition from one major environment to another (for example, from land to sea for whales; from land to air for bats) that the chances of them leaving much of a fossil record is slight. Simpson was saying that the origin of higher taxa, the major evolutionary transitions, was essentially Darwinian in character. A full series of intermediates connects ancestors with descendants. But the whole process works so rapidly, involves so few creatures, and occurs in such unusual surroundings and circumstances that we can expect to find few or even *no* intermediate fossil forms in most instances.

This absence of evidence, the lack of intermediate forms, troubled Simpson no matter how content he may have been with his explanation of that missing data. He seized upon the few specimens known (only a few have been added since) of the famous early bird *Archaeopteryx*, brandishing it to create much-needed space between his ideas and those of

Otto Schindewolf. Characterized as the prime example of *mosaic evolution* by the embryologist and evolutionary biologist Gavin DeBeer, *Archaeopteryx* is a wonderful example of just the kind of intermediate an evolutionary biologist would wish for: Far from being "halfway" between reptile and bird in all respects, it is a wonderful melange of full-fledged bird features (for example, feathers) and primitive reptilian retentions (for example, a true tail and teeth). Better yet, Simpson would have loved the relatively recent, convincing discoveries of "missing links" between archaic terrestrial mammals and true whales. These creatures, predictably enough, apparently lived along shorelines, presumably swam, but still retained four functional legs for walking on land.

To bolster his argument that major evolutionary innovation typically involves brief spurts of very rapid change, Simpson turned to another set of patterns and set up a very clever syllogism that I later dubbed "Simpson's inverse."[26] Simpson devotes considerable space in *Tempo and Mode in Evolution* to the general subject of evolutionary rates. Rather than seeing a smooth continuum from very slow to very high rates, Simpson chose to distinguish (typologically!) three distinct *classes* of rates: the extremely slow (*bradytelic*); the normal or average (*horotelic*); and the extremely fast (*tachytelic*). Most evolution falls into the horotelic category, where there is, Simpson claims, variation from moderately slow to moderately fast.

In contrast, Simpson claimed, true bradytely and tachytely are relatively rare. Because there is no paleontological evidence for tachytely, Simpson seized on bradytely—the pattern commonly, albeit loosely, known as living fossils. Simpson supposed that an accurate assessment of its causes would shed reciprocal light on the otherwise hidden, but pattern-inferred, phenomenon of tachytely. Turning to genetics, Simpson concluded that slowly evolving lineages, such as the famous horseshoe "crabs", must typically involve very large populations with relatively low mutation rates. Furthermore, natural selection will tend to keep such lineages firmly ensconced in the same, unchanging adaptive zone.

Intermediate-sized populations, Simpson thought, would be more likely to show progressive within-lineage (*phyletic*) evolution—in effect, explorations of the various adaptive possibilities latent within that adaptive zone. But it is small populations that offer the greatest opportunities for rapid evolution, even though the fate of most of them must surely be extinction. Seizing, as so many evolutionary biologists did, on Wright's pictorial metaphor of the adaptive landscape, Simpson wove a speculative scenario based on his readings of contemporary genetics, *especially* the work of Sewall Wright.

Simpson in 1944 was studying the novelties that mark the beginnings of distinctly new lineages, hence the evolutionary events surrounding the origin of higher-ranked taxa (that is, *macroevolution*[27]). To him, the problem of the origin of major new adaptations implied the actual *loss* of old adaptations before new ones could be gained. Thus his model of "Quantum Evolution" entailed three distinct phases: an inadaptive phase, where old adaptations are lost; a pre-adaptive phase, where genetic/phenotypic variation is leaning in a new direction (perhaps even fortuitously); and an adaptive phase, where natural selection grabs hold of that latent directionality in the variation and rapidly molds the new adaptive morphology.

In terms of Wright's adaptive peaks and valleys, the first, inadaptive phase sends the small population *down* the slope of the adaptive peak. This process ordinarily leads to extinction, as the adaptive valleys literally become "valleys of the shadow of death." (This is my allusion; Simpson was not so dramatic.) But some populations skate through the valley in the pre-adaptive phase, reaching the lower slopes of some other adaptive peak, high enough for selection to take over and drive the population the rest of the way up.

How are adaptations lost? What will drive a population *down* the slopes of an adaptive peak? Simpson seized on Wright's concept of genetic drift, essentially the random element working on gene frequencies within populations and acting as the foil to the deterministic process of natural selection. He also cited another statement of Wright's: In very small populations, over half of the mutations occurring are likely to be deleterious.

Simpson felt that a particular set of details of horse evolution as then known to him fit this explanatory model perfectly. Earliest horses, arising in the Eocene, were browsers. All modern horses, on the other hand, are grazers and have high-crowned molars laden with cementum to enable them to graze successfully on the glass-laden, or *siliceous*, grasses that evolved in the Miocene. For some reason, the earliest grazers had to stop browsing leaves. Simpson felt that chance variation in the dimensions of the teeth in some of these browsers, which were not optimal for browsing, accidentally set them in the direction of grazing.

As has become well-known, in 1953, when Simpson published his better-remembered book, *The Major Features of Evolution*, he modified some of his more extreme views as presented in his 1944 *Tempo and Mode in Evolution*. I remain convinced that he did so in part because Sewall Wright essentially trashed Simpson's theory of Quantum Evolution, the very core of the book, in a review published in 1945.[28] Among other things, Wright accused Simpson of errors in mathematical notation and misapplying the notion of genetic drift to paleontological pattern.

Then, too, by 1953 the pluralism that seemed genuinely to pervade the earlier stages in the development of what Julian Huxley had dubbed "the modern synthesis" in evolutionary theory was on the wane. The first synthesis was the fusion, led by Fisher, Haldane, and Wright, of genetics and the Darwinian view. This movement produced the core neo-Darwinian paradigm and saw the invention of population genetics. However, Wright and Fisher, in particular, disagreed on a number of issues, amounting to a form of pluralism in their collective efforts.

The next phase was to bring all other disciplines of biology under the general cloak of the modern synthesis. Dobzhansky extended the discourse more explicitly to the natural world and, with Mayr, added the next-higher level of discontinuity: species. Simpson brought paleontology into the fold, but he did so all the while insisting that speciation as Mayr and Dobzhansky discussed it was essentially trivial, and certainly not the stuff of major evolutionary change. The German Bernhard Rensch (like Hans Stille, the geologist encountered in the preceding chapter, working independently during World War II) also extended the neo-Darwinian paradigm to embrace large-scale patterns in evolution. And G. Ledyard Stebbins, meanwhile, brought botany into the fold.

Thus the second phase of synthesis was one of melding disparate biological disciplines into the neo-Darwinian paradigm. It could be shown that the data of each discipline generally are consistent with the new paradigm. The Society for the Study of Evolution was founded shortly after World War II (1946)—and soon thereafter a famous conference was held at Princeton in 1947 that saw the participation of geneticists, systematists, and paleontologists.[29] The conviction quickly developed that it was natural selection, as Darwin had originally said, that underlay the generation of adaptive diversity. Thus by 1953, it was more than just one negative review (painful though it undoubtedly was to Simpson)— for he modified quantum evolution to be simply a rapid burst of phyletic evolution generated completely by natural selection throughout, mirroring the zeitgeist of his times. By the Darwinian centennial in 1959, all biologists were congratulating Darwin, and themselves, on the utter triumph of the neo-Darwinian paradigm. They had, at last, forged a complete evolutionary theory. Or so they thought.

Thus Simpson's original notion of quantum evolution was forgotten, put to rest by its very own author after suffering a broadside from Wright and in consideration of the general stance on selection adopted by the biological community at large. True, Simpson's imagery of peak-hopping as the central metaphor for adaptive evolutionary change, based on Wright's work, has been resurrected to some degree by modern geneticists.[30] But his original scheme was, in fact, based on too many dubious propositions: There really is no reason to believe that evolutionary rates are distributed into three distinct classes, each with its own distributions. Or that these three putative categories of rates correspond to the three equally distinct "modes" of evolution: speciation, phyletic evolution, and quantum evolution. Or that adaptive change necessitates the loss of prior adaptations before new ones are gained. Or that species themselves in fact occupy adaptive zones or peaks, meaning ecological niches.

Thus, the actual content of the original formulation of quantum evolution is not something worth remembering. However, the very nature of Simpson's enterprise *is*. For he finally brought the spirit of James Hutton into the arena of evolutionary biology. Simpson looked at patterns that he saw in the rock record of life's history and said, in effect, that these apparent bursts of rapid evolutionary change are *not* the artifacts of a poor or poorly studied fossil record. They are too pervasive to be wholly artifacts. They must, instead, be telling us something about the nature of the evolutionary process itself. These abrupt origins of major new lineages must be *real*. It is rapid evolution that produces the gaps, not hiatuses in sedimentation.

And Simpson went further. He turned to genetics, in what he correctly saw as a new field, rich with understanding. He examined what he called the *evolutionary determinants*: all of the processes and parameters in the purview of genetics. He reviewed each determinant in his second and longest chapter of *Tempo and Mode*: variability of organisms within populations, the nature and rate of mutation, generation time, population size, and natural selection. Unlike some of his predecessors from the field of paleontology, he made no attempt to add to the list, to suggest other factors that geneticists themselves hadn't considered. What he did, instead, was to say that the patterns he was addressing in the fossil record call for a new theory, a novel concatenation of the determinants. In short, novel causal theory.

The importance of this bold and imaginative gambit can hardly be overemphasized. A century and half earlier, Hutton had looked at patterns in the earth's crust and from the magnitude of the effects deduced the existence of processes such as mountain building, and granitization, which had not been suspected before. He inferred that heat flow from the interior of the earth must be implicated in these processes. Hutton essentially founded geology. The irony of his work is that he spawned a literal dichotomy, where one branch of the field ran off to study the processes and the other decided to unravel the history wrought by those processes. Not until plate tectonics came along has there again come to be a form of resonance, of mutual illumination, between the events of earth history and the study of its dynamical processes.

Simpson, a young evolutionary-minded paleontologist, founded no new field. On the contrary, he came along three-quarters of a century *after* Darwin had established evolutionary biology. He wasn't even the first paleontologist to address evolutionary issues. But he was the first to establish in mainstream scientific circles the notion that evolutionary theory can in principle—in fact, *must*—be held accountable to real patterns in the history of life. Thus Simpson was not resurrecting some time-honored behavior that his contemporaries had lost sight of. Rather, like Hutton, he was starting something new, though, unlike Hutton, it was many years after the field had become established. Simpson taught us to take *what we think we know* about the mechanics of generation-by-generation changes (and stability) in gene frequencies and to compare that with evolutionary patterns *we actually see* in the fossil record. He taught us to be bold in forging new combinations of familiar processes to forge a better fit between what we think is going on and what we see in the record of the rocks. That's how punctuated equilibria came into being.

Punctuated Equilibria

The Darwinian centennial of 1959 seemed to its celebrants an occasion to proclaim the end of dissension. At last a full, or reasonably complete, evolutionary theory had been achieved, or so it was commonly claimed. Gone, or swept into the deep background, were some of the more egregiously unusual elements of the early synthesis. Both Wright's notion of genetic drift *and* Simpson's use of it (or mishandling, according to Wright) to formulate aspects of quantum evolution had long since left center stage. All was seen to be virtually linear, adaptive change within lineages: Darwin's simple original vision, now thoroughly melded with genetics.

Yet 1959 was really more of a watershed than a true completion of evolutionary theory. Indeed, the gap immediately began to widen between traditionally estranged camps. In one direction lay the strong movement back to Fisherian first principles, with its core postulate that natural selection is not only necessary, but entirely sufficient, to create the evolutionary history of life. This movement led to ultra-Darwinism, which was spearheaded by the appearance in 1966 of George Williams's estimable book *Adaptation and Natural Selection*.

On the other hand, a separate movement, coming from Columbia University's Department of Geology in the mid-1960s, worked instead to effect a conceptual fusion between the Dobzhansky-Mayr views of the nature and evolution of species (whose importance was explicitly denied by Simpson) and Simpson's Hutton-like use of evolutionary, historical pattern to elucidate aspects of biological evolutionary process. In many ways, such a fusion is precisely what punctuated equilibria *is*.[31]

I have already recounted in Chapter 1 four basic patterns that led to my early development of the core of punctuated equilibria. Physical anatomical patterns in trilobite eyes showed growth, then stability, within population samples—but some variation in that stable number between populations. When those patterns were mapped, a larger-scale geographic pattern popped out, one that looked like Mayr's patterns of geographic variation. When maps were compared at different intervals in a 6- to 8-million-year sequence (of Middle Devonian time in eastern and central North America), what looked like true geographic, or allopatric, speciation emerged.

The only real difference lay in one additional pattern not discussed by Dobzhansky, Mayr, Wright, or even Simpson: stasis, the entrenched stability that characterizes the vast bulk of most species' histories.[32] Stasis is an empirical pattern. As we've seen, it was well-known to Darwin's paleontological contemporaries. But in the early decades after 1859, stasis had quickly become pattern denied, so contrary to Darwinian expectations did it seem.

Thus we needed new theory to explain the sorts of patterns seen in my *Phacops rana* lineage data. Insofar as patterns of geographic variation and speciation were concerned, this was simply a matter of exchanging one set of theoretical expectations for another. Thus, the slow, steady, progressive, *phyletic* evolutionary change under the assumed guidance of natural selection was exchanged for Mayr's allopatric model of speciation. Rather than natural selection simply modifying species lineages linearly through time, other evolutionary principles (geographic variation and speciation, along with natural selection) were far more apt. And they provided a far better "fit" to the data.

I tend to represent myself as a "knee-jerk neo-Darwinian," someone who embraces adaptation through natural selection as a thoroughly established cornerstone of evolutionary biology. In applying geographic variation and speciation as interpretive principles to my paleontological patterns, I was being utterly conventional. (The only criticism that the speciational aspect of punctuated equilibria has ever received is that it is not particularly "original."[33]) Yet what is important is that the patterns themselves *forced* the interpretational switch from Darwinian (and neo-

Fisherian) emphasis on simple adaptive transformation within lineages to one emphasizing the origin of discontinuities among far-flung populations of ancestral species. There is no point to inventing theory where an existing alternative will do quite nicely.

Stasis has been another matter. Over the years, Gould and I have explored a number of possibilities for why it happens (which has led to charges that we are forever changing our minds). One early suggestion was made in our original 1972 paper naming punctuated equilibria and fleshing out its details and some of its implications. Here we invoked biologist Michael Lerner's then-fashionable *genetic homeostasis*. This notion involves a sort of intrinsic, ingrained genomic conservatism, and it stresses the difficulties inherent in modifying the genetic basis of developmental (embryological) systems. However, genetic homeostasis never gained much popularity, and we find ourselves still criticized for even making such a suggestion.

Far closer to the mark, however, is one of my preferred explanations of stasis. It is one that says, paradoxically (and contrary to at least superficial Darwinian expectations), that *stabilizing natural selection will be the norm even as environmental conditions change*—so long, that is, as species are free to relocate and "track" the familiar habitats to which they are already adapted. Rather than remaining in a single place and adapting to changing conditions, species move. And so they tend to remain more or less the same even if the environment keeps on changing.

This may sound like novel theory but really is just a modification of older common sense to embrace a pattern that biologists and paleontologists have documented in abundance in the past few decades. Cardinals, tufted titmice, red-bellied woodpeckers, and mockingbirds are now common backyard denizens in the northeastern United States. They all have moved up north, along with opossums and many other species, marine as well as terrestrial, as the climate has warmed considerably over the past century. There is nothing mystical or mysterious about this. Habitat tracking is the reason why there were lions and hippos in Trafalgar Square during a very warm interglacial period about 120,000 years ago. Yet the mere suggestion of habitat tracking as the explanation of stasis has provoked howls of protest from some quarters.[34] Are paleontologists not supposed to dream up "new" mechanisms, even if they are merely the application, once again, of thoroughly observed and understood common natural patterns?

However the work of my colleague Bruce Lieberman has convinced me that it is an idea latent in Sewall Wright's very conception of the internal structure of species that has the most to do with the great stability shown by virtually all species ever encountered in the fossil record.

Lieberman has found a very telling pattern: If you study population samples of a species drawn from a variety of different environmental settings, those samples show more change, more phyletic evolution, *within* the same environments than the species as a whole does when all samples from all environments are lumped together. Habitat tracking, at least for populations within a species, seems to allow for some degree of adaptive change. But that change is typically masked, muted, or lost when applied to the evolutionary history of the species as a whole.

In short, each local population of a far-flung species will be living in ecosystems with somewhat different physical environments, predators, and prey. Moreover, each local population will have its own sampling of the genetic variation of the entire species, as well as different mutational histories. It will also have a different history of genetic drift and, critically, natural selection than that experienced by local populations of the same species living in different ecosystems. In short, it is highly unlikely that natural selection could ever "move" all the populations of an entire species in any one single evolutionary direction for any significant amount of time at all. Lieberman seems to have confirmed Wright's unspoken prediction: Stasis is implicit in the very structure of species in the wild.

The entire process of using historical, paleontological pattern to make modifications of evolutionary theory is driven mostly by the invocation of different sets or combinations of explanatory theory than had previously been invoked. This is just what Simpson, in his Huttonian mode, taught us to do. That even such relatively benign behavior sparks occasional protests from the guardians of that explanatory theory really should not matter. In this instance there has been, and will be, no retreat.

But consider a further gambit, one in which historical pattern has led us to propose an actual *additional theory*. This new one has analogies to existing theory (natural selection, to be precise), but it nonetheless is novel. In 1972, Gould and I attempted to explain how species in an evolving lineage can show directional change through time *despite the fact that each species itself remains relatively stable throughout its history*. We suggested that a higher-level process, variously called *species selection* or *species sorting*,[35] may be at work. This process imparts directionality between species otherwise missing from the stasis-laden histories within species.

Radical as the above suggestion may seem, it too is based on empirical paleontological pattern. At the heart of the matter lies the inescapable fact that species have births (via speciation), histories (typically, if not invariably, somewhat stable), and deaths (that is, extinction). Species are spatiotemporally discrete, historical entities. They are not, the record suggests, simply the arbitrarily recognized subdivisions of smoothly and continuously evolving species-lineages.

Well, if species are individuated, why not imagine that there will be patterns of differential births and deaths of component species within a multispecies lineage? And indeed, that's precisely what the record shows, in records of African antelopes, of humans (documenting the brain's size increase over the past 4 million years), or just about any lineage where clear-cut evolutionary trends accumulate through time.

Simpson was right: The record has patterns fraught with evolutionary theoretical meaning. He was wrong, of course, in dismissing species as trivial. They are, instead, major players. So-called higher taxa such as genera are simply ancestral-descendant strings of species. They are products of history possessing no true internal dynamics other than the pattern of genealogical ancestry and descent that binds them together.

Species, in contrast, do have an internal dynamic: the ongoing reproductive behavior of component organisms in their semiseparate populations (*demes*). And species, like organisms, *do* something: They can make more of themselves (they *speciate*). The fragmentation of reproductive communities that *is* the process of speciation amounts to (re-)packaging genetic information. Speciation fragments gene pools, giving independence to new sets of genetic information, inaugurating their own independent histories. It is the fates of independent species, more than anything else, that constitute macroevolution, the large-scale patterns of evolutionary history.

Thus punctuated equilibria, with all its extensions and ramifications, is built on the twin, solid bases of Mayr and Dobzhansky's concern with species-level discontinuity and the process of speciation, and the Simpson concern that evolutionary theory be held accountable to actual (repeated and well-documented) historical pattern.

Above all, punctuated equilibria takes the embryonic discussion of hierarchy in the earliest pages of Dobzhansky's first edition of *Genetics and the Origin of Species* (1937), and develops that concept more fully. For Dobzhansky truly is the father of the what for the last 15 years I have been calling the genealogical hierarchy: genes of the germ-line replicating, making more of themselves; organisms reproducing, making more of themselves; demes coming and going, in so doing making more of themselves; and species speciating, making more of themselves. This genealogical hierarchy is a hierarchy of genetic information, one in which the individual components of any given level are held together by the reproductive ("more-making") activities of the individual components of the next level down. Species (and demes) are held together by ongoing reproduction of organisms, who themselves owe their existence to the continual production of eggs and sperm. And, back on the upper side of the equation, higher taxa, strings of ancestral-descendant species, are themselves held together by the speciating propensities of species themselves.

The genealogical hierarchy is very real. It is an abstraction, really, of all the patterns emphasized by evolutionary biologists since Darwin: patterns of differential birth and survival, or reproductive success (that is, natural selection); patterns of speciation; patterns of differential species success.

But powerful as the genealogical hierarchy is, and however central its hierarchically structured packaging of genetic information is to the very definition of the term *evolution*, something obviously still is missing. For there is no formal, explicit connection, in either these abstractions or even in the empirical historical patterns on which they are based, with the world of matter-in-motion, or matter-energy transfer. I have acknowledged several moves to incorporate these concepts into the general Darwinian genealogical paradigm. Examples of such moves are the abortive Dobzhansky/Mayr (and even Simpson) attribution of ecological niche attributes to species, or the general rubric of co-evolution, where adaptive changes in one species are in response to aspects of stasis and change in other species. Indeed, most adaptations are not about reproduction, but rather concern energetics—obtaining food, for example.

Such are all hints at the nature of the links connecting biological evolution to principles of matter-energy transfer. It is time now to flesh these links out and make them explicit, to connect evolution with the rest of the physical realm once and for all.

CHAPTER 6

The *Cecropia* Moral:
Evolution and the Realm
of Matter-in-Motion

We still have a way to go along the *Cecropia* trail. We still need to discover the links that connect the realm of physical processes which, on various scales, shape the earth and its history with the genealogical process of descent-with-modification: biological evolution. What we seek is something lawlike, something that derives its substance from repeated, hauntingly similar events linking earth and life: linked patterns of biological evolution and earth history.

Within the purely genealogical realm, the *Cecropia* trail has taken us as far as we can go. To go farther we must leave pure genealogy and look to the dynamic mixes of species that form the small and large systems of biological matter-energy transfer: *ecosystems*. This much we already know from the trail that began in Chapter 1: Organisms may be on a quest to reproduce, but even more they are quintessentially matter-energy transfer machines. Even the simplest single-celled organisms need energy and nutrients just to exist. Multicellular organisms additionally require energy and nutrients to develop from a fertilized egg, to grow, and to maintain the body until eventual death. Reproduction itself also depends, ultimately, on such energy input.

This is the realm traditionally given over to ecology. But even though evolutionary biologists (as well as evolution-minded ecologists) have always claimed to have taken this connection seriously, little explicit attention has been paid to this connection in the annals of evolutionary theory—or in ecological theory, for that matter. Evolutionary biology's prime focus has been on Darwin's concept of adaptation through natural selection, where most adaptations are concerned with energy procurement, utilization, or the avoidance of serving as an energy source for some other organism. But beyond that, evolutionary biology has remained oddly incurious about the precise nature of the links that must exist between evolution and the physical universe.[1]

What is at stake here? Nothing less than the answer to the question: What drives evolution? Ultra-Darwinists look internally. They claim that competition for reproductive success among genes, or at least among organisms of the same species within local populations, is the primary driving force. From this perspective, environmental change merely signals a shift in the boundary conditions, the details of the game of reproductive success.

Paleobiologists, on the other hand, are increasingly convinced that the basic patterns in the history of life are both episodic and characteristically cross-genealogical, meaning that evolutionary pulses typically affect many lineages of a region simultaneously. Somehow the ecological realm must be more explicitly linked to the evolutionary process than evolutionary biologists have been willing to admit—or at least specify.

It seems clear to me, then, as it did to Simpson in the late 1930s, that the paleobiological and the deeply reductionist genetical views on evolution must both be "true" in principle.[2] But let's note now that nothing in the ultra-Darwinian argument prepares a biologist for the conclusion that evolutionary history is overwhelmingly episodic and cross-genealogical, or the generalization that most evolutionary change consistently appears to have been triggered by physical events in earth history. If common ground is to be sought, at the very least additional theoretical structure must be developed. Such a structure must go beyond the genetics of organisms within populations, or even beyond concepts such as punctuated equilibria, which acknowledges the episodic, but not cross-genealogical, aspects of patterns in the history of life. We will need, in short, to add *ecologically based theory* beyond Dobzhansky's genealogical hierarchy.

But first, let's look for more patterns, ones that establish beyond any doubt the validity of seeing *physical* events as the main impetus for evolutionary stasis and change over the past 3.5 billion years.

Links to Matter-in-Motion

In general, and on average, there are more species in both terrestrial and marine habitats in the tropics than there are in polar regions. In general, this pattern is monotonic, or smoothly progressive. It progresses from ecosystems with relatively few species in the extreme higher latitudes to ones with the most species in the tropics.

This latitudinal diversity gradient is a well-known, long-appreciated biogeographic and ecological pattern. Two features of this pattern are of the utmost importance. First, by its very nature the pattern is correlated with, suggesting it maybe causally linked to, the amount of sunlight reaching the earth's surface. Second, the pattern is more than purely ecological; it is also genealogical, or evolutionary, in nature. With notable exceptions,[3] the pattern holds for phylogenetic group after group. The preponderance of tropical biota is thus a large-scale pattern that is linked in a cross-genealogical, ecological fashion with both ambient solar radiation *and* the evolutionary history of many distinct taxa of organisms. A third feature of the pattern cements this dual relationship still further. It has been in place throughout the history of life, or at least since the advent of a rich metazoan fauna, and therefore a dense, pattern-preserving, and analyzable fossil record.[4]

What is the underlying cause of the latitudinal diversity gradient? Answers differ, depending on the eye of the beholder.

An ecologist, while acknowledging the evolutionary component of the pattern, tends to look for explanations strictly in terms of moment-by-moment ecosystem dynamics, patterns of matter-energy transfer linking populations into coherent, local ecosystems. Why are tropical ecosystems typically more diverse than higher-latitude systems? Because, an ecologist is prone to answer, tropical habitats are typically finer-grained, as there are more potential "niches" to be filled. But then why are there more niches in the tropics than in the higher latitudes?[5]

An evolutionary biologist, on the other hand, is more likely to start with a different set of questions: Are the tropics a "diversity pump," churning out more species per unit of time than higher-latitude environments? Or is it a matter of the tropics acting like a "museum" where species diversity is differentially preserved, not generated? Or is it in some fashion a combination of the two? And, again, the underlying question: Why? What would make the tropics either a diversity pump or a museum?

A quick survey of the thinking about this problem reveals much common ground between the two initially disparate points of departure—ecological and evolutionary. This is not to say that ecology and evolutionary biology are effectively melded after all. But a general set of problems, a fundamental pattern, a *fact*, of life with obvious ecological and evolutionary components exists for which, unsurprisingly, a number of evolutionarily minded ecologists and ecologically minded evolutionists have sought a general explanation.

My favorite explanation, as a good illustration of combining ecological and evolutionary theory to explain a fundamental pattern of the living world, is as follows[6]: The amount of solar radiation reaching the earth is very different in the tropics than in the higher latitudes. Yet it is not so much the total amount of sunlight as its patterns of distribution that appear to underlay the diversity gradient. Specifically, daily and seasonal fluctuations vary greatly from the tropics to the poles. On the equator, there are 12 hours of daylight, 12 of night, and virtually no seasonal temperature fluctuations. On the poles, there are 6 months of continuous daylight and 6 months of perpetual night, though solar radiation waxes and wanes during the 6 months of continual daylight.

Consider the midrange, the temperate zones with distinct seasonality. Trees shed their leaves and grasses turn brown as photosynthetic factories are turned off for the winter because water-sap transportation systems cannot operate under freezing conditions. Some birds and insects (monarch butterflies, for example) migrate southward to warmer climes. Other insects, reptiles, amphibians, and mammals hibernate. Still other species, such as woodpeckers and many seed-eating birds, tough it out. They either move southward just a bit or stay put on their breeding grounds. All species, in other words, have some strategy for dealing with shifts in temperature and food supplies.

Thus higher-latitude species are physiologically able to withstand pronounced seasonal and daily climatic fluctuations, especially temperature and its secondary effects. Not surprisingly, such physiologically generalized organisms can occupy a broad range of habitats. So the high-temperate and subarctic species, such as moose, wolves, red foxes, and mallard ducks, typically occur worldwide. The late Roger Tory Peterson was able to cover in just two volumes all North American birds—eastern and western. He found the dividing line for the two regions to be just east of the Rock Mountain front. And many species such as the bald eagle and American robin were found to occur continent-wide. Temperate species tend to be broad-niched and wide ranging.

Not so tropical species, which do not experience such diurnal and seasonal temperature fluctuations. There is far less reason for organisms to be physiologically generalized in the tropics than in the higher latitudes. Natural selection, then, would be expected to narrow adaptations to finer subdivisions of habitat. With narrower adaptations, reflecting narrower niches, we would also expect smaller geographic ranges of tropical species compared to higher-latitude species. And this we do find as a rule.

Restating the comparison: Broadly niched species cover a lot of ecological bases. They live in many habitats and consume a wide variety of resources. More narrowly niched species, in contrast, are focused on a smaller range of resources. In other words, the ecological pie is divided up much more finely in the tropics than in the higher latitudes. The result is that less evolutionary change is required for a new species to establish an ecological footing in the tropics than in the higher latitudes.

And the conclusion is supported by yet another pattern: Some groups with high species diversity in the tropics (for example, tropical American flycatchers, wrens, and woodcreepers) nonetheless show relatively little anatomical diversity. Many species in the same group look confusingly alike, a sign that their populations are occupying none-too-dissimilar niches.

There are further consequences. Paleobiologists have shown that global mass extinction events have disproportionately high effects on the tropics. A higher percentage of tropical species become extinct in these spasms, especially those attributed to episodes of global cooling. This occurrence is generally interpreted to reflect the narrower-niched adaptations of tropical species.

So I prefer the explanation that the tropics are a peculiar hybrid of diversity pump and museum. There is no evidence to show that reproductive isolation (that is, true speciation) occurs more frequently in the tropics than elsewhere. But, because less evolutionary change is required for new niches to be occupied, fledgling species have a higher evident rate of survival in the tropics. Thus, the tropics invariably accrue species at a faster rate.

What counts here is not the success of this particular explanatory scenario, but its form: What seem to ecologists to be large-scale ecological or biogeographical patterns may seem to evolutionists to be purely evolutionary patterns. Because the global latitudinal diversity gradient patently displays elements of both worlds, such patterns, and their explanations, hint at what a general theory linking evolutionary history with ecological process might look like.

Perturbations of these patterns, particularly in ecological systems, seem to lie at the heart of most evolutionary changes in the history of life. Easiest to see are the evolutionary consequences of global mass extinctions. Many extinction events have punctuated and stimulated the history of life since the advent of complex animal life some 540 million years ago. Five in particular stand out as truly global in scope. They affected nearly all marine and terrestrial ecosystems, and claimed species from virtually all documentable lineages extant at that time.[7]

As we discovered in Chapter 2, Baron Cuvier saw such events as extinction catastrophes followed by separate acts of divine creation to repopulate the planet. One might reasonably have expected new theory to develop once Darwin had successfully overthrown the idea that biological diversity reflects the special act of a supernatural Creator. In other words, once Darwin had convinced the thinking world that life has had a natural, evolutionary history, one would have expected biologists to reinterpret the abrupt events of biotic turnover as extinction events followed by bursts of evolutionary diversification. No such thing happened. Biologists have ignored even the most glaring, blatant, and general patterns in the history of life. From the mid-nineteenth century right up to the present day, they have preferred instead to elaborate theory pertaining to single lineages. Indeed, the most recent focus has been on populations, not even entire species or, still less, species-lineages. They have offered these explanations without explicit regard for characteristic kinds of physical events affecting the planet.

Yet pattern cannot forever be denied. Beginning with the work of paleontologist Norman D. Newell,[8] who came to the American Museum of Natural History in New York right after World War II, paleontologists have been rediscovering the rich, physically caused, cross-genealogical patterns that are so typical of the evolutionary history of life. As a result, when paleontologists these days address the broader aspects of the history of life, what they have to say is radically different from what they were saying, say, in the Darwinian centennial year of 1959. Back then, the history of life was seen more or less as a smooth continuum, broadly, if only "in principle," reducible to the core neo-Darwinian paradigm of progressive adaptive change through natural selection. Nowadays, the rhetoric is very different. Most paleobiologists say that nothing very much happens in evolution unless and until physical events overwhelm ecosystems, whether locally, regionally, or globally. Thereby extinctions are triggered and the evolutionary clock is reset.

Most dramatic, and perhaps most easily read, are the evolutionary consequences of the great global mass extinctions. Typically, the more global and all-encompassing the event, the longer the extinction takes to unfold. (The end-of-Cretaceous meteor event is an apparent exception.) And the longer, too, it takes for sufficient evolution, the production of new species, such that ecosystems regain a look of "normalcy" to them.

Perhaps the most graphic example of how differently the history of life appears to paleontological eyes at the twentieth century's end compared with midcentury perspectives is the contwined fates of dinosaurs and mammals. Traditional perspective had it that the extinction of dinosaurs was inevitable. Recall that Darwin argued that extinction was integral to the evolutionary process. But he thought it was so only in the sense that superiorly adapted species would outcompete less-advanced ones, thus driving them to extinction. Taxonomic gaps were simply that, gaps, species falling by the competitive wayside. (Recall that Dobzhansky, too, felt that gaps are intrinsic to the evolutionary process. But he thought they were produced, not by the extinction of intervening species, but by the process of speciation itself.) Mammals were held to be simply too intelligent and, with placental development (which had evolved long before the end of the Mesozoic), reproductively more efficient as well. Thus, it was generally agreed, it had been in the cards all along that the "superior" mammals would eventually displace dinosaurs, outcompeting them and literally wresting their ecosystemic roles away from them.

Paleontologists in Darwin's day knew of both dinosaurs and mammals in Upper Triassic rocks (some 220 or so million years old). Extensive collecting for the last century and a half hasn't changed that picture. Mammals and dinosaurs apparently both did arise roughly at the same time during the Upper Triassic. Their appearance followed the demise of the rhynchosaurs, the then-dominant vertebrates of the non-Gondwanan continents.[9] For reasons no one pretends to understand, it was the dinosaurs, not the mammals, that radiated, evolving into an array of large and small herbivores, carnivores, and scavengers. Reptilian collateral kin (pterosaurs, ichthyosaurs, mosasaurs, and plesiosaurs) were the tetrapod vertebrates of the air and marine- and freshwaters.

Dinosaurian ranks were occasionally pruned in extinction events, including the great one at the end of the Triassic which depleted their ranks severely. But some dinosaur species managed to survive each such event. After a lag, they evolved a new spectrum of species that formed the terrestrial vertebrate component of the newly restored ecosystems. For example, by the end of the Triassic, dinosaurs appeared in the ecosystems of Gondwana after the extinction of the dominant mammallike reptiles in the southern-hemisphere continents.

But the fateful day eventually did come, not when the mammals rose up to claim their rightful control of the world's ecosystems, but when the dinosaurs (who had been dwindling in any case) finally succumbed utterly. In the violent scene that put an end to the Mesozoic world, the "Age of Reptiles," every last species was extinguished. It was only then that mammals, in the absence of their ecological rivals whose very presence had kept them as fairly generalized, smallish, and undifferentiated physically, suddenly blossomed forth (after a characteristic lag period). Now, after 150 million years, it was the mammals who diversified into herbivores, carnivores, and scavengers of a wide variety of shapes and sizes.

In brief, the only thing "inevitable" about the evolutionary diversification of mammals was that dinosaurs could not last forever. This was not because of any innate inferiority, but because of the long-term inevitability of large-magnitude physical events, ones in which adaptations of organisms and entire ecosystems are simply overwhelmed. Survival is as much a matter of luck as anything else. And evolution rebuilds based on what survives.

For the last half-billion years following the advent of large, multicellular forms of life, the most fundamental divisions of the geological timescale are marked by these major episodes of mass extinction. The Permo-Triassic "Great Dying" marks the end of the Paleozoic Era and the beginning of the Mesozoic. The demise of the dinosaurs and so many other plants and animals at the end of the Cretaceous equates to the end of the Mesozoic and the beginning of the Cenozoic Era (the "Age of Mammals"). Yet the geological timescale is an almost infinite regress of subdivisions within divisions: Geological periods divide up the eras, while periods themselves are divided into epochs. Each of these progressively finer sets of subdivided time corresponds to bodies of rock containing patterns of physical events, many of which, like the fall and rise of sea level, often seem almost hypnotically rhythmic. And nearly all these events had an effect on life, changing the distributions of habitats, disrupting ecosystems and, when boundaries of tolerance were transcended, driving species to extinction and eventually setting the stage for more evolution to occur. The more global and taxonomically all-encompassing an extinction event is, the longer the lag period (up to 5 million years, or perhaps even longer, in the case of global mass extinctions). But so far these physical events have not been able to extinguish life altogether. Sooner or later natural selection and speciation, modifying and diversifying the remnants of the previous biota, begin the process of recovery. The greater the extinction event, the more profound the change in composition of those ecosystems and the longer it takes for their components to evolve.

The aftermath of these global mass extinctions may be the most dramatic and easiest to read, but the dynamics of the evolutionary process are bound up with populations and species. It is the more local, less all-encompassing or devastating ecosystemic disruptions and species extinction spasms that reveal the details on how the evolutionary world links up with the realm of matter-in-motion. Natural selection acts within local populations having niches in local ecosystems. As Darwin and Wallace both realized, the "selectivity" of natural selection occurs only because population numbers are limited. The bounds are set, in large measure, by the carrying capacity and other "limiting" factors (such as disease and predation) intrinsic to a population's niche in the local ecosystem.[10] Hence, as already remarked, most adaptations configured by natural selection pertain to the economic life of individual organisms. This is the most solid link between evolution and the world of matter-in-motion as yet fully specified by evolutionary theory.

Yet this link is clearly not enough. It does not tell us why stasis is so prevalent, or how it is that most adaptive change appears to be corre-

lated with speciation events. Nor does it tell us in general why speciation itself occurs when it does, and why speciation is relatively rare during the course of the long history typical of individual species. Here is
where both ancient and recent cross-genealogical patterns of extinction
and speciation are connected with the stability and disruption of ecosystems. And this sheds much-needed light on how ecology, the biological aspect of matter-in-motion, is linked with evolutionary history.

Disturbance, Stability, and Evolution in Ecosystems

As we've just seen, life is incredibly resilient. This resilience is a hallmark of life at all conceivable ecological levels. Mass extinctions are
patterns at the high end of the scale. As already noted (see note 7), all
the evolutionary diversity required to rebuild the world's ecosystems in
the Lower Triassic apparently came from as few as 4 percent of the genetic variation that survived the Permo-Triassic crisis. But much the
same resilience shows up on smaller scales as well. So to get to the core
dynamics of stasis and evolution, we must look at finer-scale historical
patterns.

Individual organisms are adapted to the vicissitudes of life: the daily
and seasonal variations of rainfall and temperature, of food and sometimes oxygen supply, of predation and disease. For example, when extreme low tides expose the mussels in an intertidal zone for
longer-than-average periods of time (24 hours or more), once in a very
great while, the vast majority survive by literally clamming up and shut-
ting down. They do so until the food- and oxygen-bearing seawaters are
finally restored.

Much the same is true of ecosystems. When local ecosystems are significantly degraded, the populations of most, even all, component species
are totally wiped out. Yet, in due course, in a matter of years, the system is
rebuilt in more-or-less recognizable form. Such is the lawlike process of
ecological succession. This is the moral of the *Cecropia* stands on the
flanks of El Yunque in Puerto Rico. After most of the forest was blown
down by Hurricane Hugo, weedy pioneer species, such as *Cecropia*, invaded the destroyed area. These trees started a forest that, in a short time,
will have only a few isolated pioneers like *Cecropia*; the rest will consist of
the same species as were present before destruction rained on the area.

In ecological succession, replacements are recruited from the outside, from surviving populations who are able to disperse to the area as
seeds, marine larvae, or, in the case of organisms like mammals and
birds, as adults. (The *Cecropia*, recall, cheats a bit by leaving its seeds in

a dormant state all over the place, ready to take over when opportunity knocks.) This is similar to yet a larger-scale kind of resiliency already encountered in Chapter 5: habitat tracking—cited as a partial explanation for the common pattern known as stasis.

What happens when environments change? Specifically, what happens when global average temperatures decline 10 to 15 degrees centigrade? The traditional assumption is that natural selection will modify the adaptations of species, provided the needed genetic variation is present, so that the species will survive. What really happens is well known to all ecologists, and is especially clear when it comes to climatic change: species move. So long as organisms can "recognize" suitable habitat—that is, conditions to which their evolutionary adaptations are already suited—and so long as they can get there, species will survive, and they will survive relatively unchanged. Even stationary species such as rooted corals, barnacles, and sponges in the oceans, and plants on land, routinely engage in habitat tracking.

Thus, species stasis and species survival are two separable, though clearly linked, results of habitat tracking.[11] Extinction comes if environmental change is too rapid, or if recognizable habitat elsewhere simply cannot be located. But ecologists discussing patterns of ecosystem fragmentation and reassembly are unsure whether the ecosystems themselves exhibit long-term stability or whether, as the data actually seem to suggest, they are reassembled in somewhat different form as climate shifts around. A good example of this is the total pattern of species movement during the Pleistocene, or "Ice Age," when four major pulses of global cooling, each with many minor oscillations, brought massive ice fields from the arctic down into today's temperate midlatitudes of both North America and Eurasia. In other words, the composition of plant, animal, fungal, and microbial species composing ecosystems changes as climate shifts. Not all species move equally quickly, nor will all species find their new terrain (with variant chemistries and topographies) equally suitable. Put another way, every species has local populations living over a somewhat varied set of ecosystems; generally, no two species (whether closely related or not) will have precisely the same geographic range, which all boils down to no two species likely to find equally "suitable" newly pioneered territories.

Ecologists are prone to debate how tightly integrated ecosystems tend to be—in other words, how strongly interdependent the particular mix of species are that form the biological part of an ecosystem.[12] There are plenty of examples of large-scale biotic change occurring if just one "keystone" species is removed from an ecosystem.[13] But historical patterns of mix-and-match habitat tracking suggest that most ecosystems can tolerate the presence or absence of quite a number of the components found in otherwise very similar surroundings elsewhere.

Enter Evolution

There are clear connections between these varying ecological patterns of resiliency, from the smallest scale of the individual organism, through ecological succession, to the even larger scale of habitat tracking. Individual organisms and, in the latter two cases, entire species tend to survive by moving around, sending out propagules to rebuild ecosystems, whether locally degraded (*Cecropia* on El Yunque) or regionally revamped (as when glaciers slowly move south from the arctic). But evolution is classically about *change*. So far, local and regional patterns of ecological resiliency imply stability of individual species lineages, not evolutionary change. Where and how does real evolution come into this picture?

Consider the effect of Hurricane Hugo on El Yunque, and on the entire island of Puerto Rico, for that matter. Prior to Hugo's hit in 1989, the endemic Puerto Rican parrot had been reduced to fewer than 100 known individuals living in the Loquillo Mountains, of which El Yunque is one. Agriculture and urbanization had already transformed so much of this bird species' habitat that it was on the verge of extinction. Hugo took about 50 percent of the remaining birds. Though the population has since recovered to approximately pre-Hugo proportions, and is now being augmented by a captive breeding program, Hugo might very well have done away with these beautiful animals entirely.

In other words, physically induced ecological calamity, if great enough in areal scope and intensity, can drive *all* the populations of a species extinct. Indeed, it can drive *many different species* extinct all at the same time. And that's exactly what we paleontologists see in the fossil record as the dominant pattern, not only of extinction, but of evolution as well.

It is not just single species that are in stasis. Virtually all the component species of regional ecosystems are evolutionarily stable, often for millions of years. Of course, that's only half of the pattern. Periodically, the majority of these species disappear, to be replaced, in due course, by others. One way of looking at this pattern is to see it as the ecological generalization of stasis and change that underlies the notion of punctuated equilibria. It is a simple, ineluctable truth that virtually all members of a biota remain basically stable, with minor fluctuations, throughout their durations. (Remember, by "biota" we mean the commonly preserved plants and animals of a particular geological interval, which occupy regions often as large as Roger Tory Peterson's "eastern" region of North American birds.) And when these systems change—when the older species disappear, and new ones take their place—the change happens relatively abruptly and in lockstep fashion. It affects most of the species of a region more or less at the same time. Evolution goes hand in hand with the degradation and rebuilding of ecosystems, and the origin of new species depends in large measure on the extinction of older species.

The fossils I know best are from the eastern United States, occurring in rocks that record a 6- to 8-million-year segment of time of the Middle Devonian Period (beginning roughly 380 million years ago). Paleobiologists Carlton Brett (from the University of Cincinnati) and Gordon Baird (from the State University of New York at Brockport), joined by me and numerous students, have attempted to integrate the Brett-Baird picture of ecological structure and history of this so-called Hamilton fauna with evolutionary patterns of stasis and change of several hundred known species.

Brett and Baird have identified some eight recurring assemblages of common invertebrate species.[14] These assemblages have characteristic numerical dominance of the same sets of species, repeated over and over again. There are occasional fluctuations: Some species are relatively more numerous in some places than others, some are absent altogether in some samples. But the degree to which this inconsistency reflects actual variation in community structure or, rather, a reflection of sampling problems (notorious with the fossil record in any case) cannot be said.

As time went on and mountains rose in what is now eastern New York State and further south, great deltas spread, in pulses, across the region to the west. (The Catskill and Pocono "mountains" are remnants of the easternmost of these deltaic complexes.) The seafloor communities tracked those shifting shorelines, oscillating first westward, then back eastward, with an overall encroachment to the west. I don't suppose that the communities always shifted together in one continuous stream. Rather, it is more likely that, episodically, conditions arose suitable for the various different associations. The largely planktonic (that is, free-floating) larvae of the component species probably wafted around in the current, found these new habitable zones, and settled down and founded new communities.

And thus did life go on for some 6 to 8 million years. Nothing much happened. At least one common species disappeared halfway through the interval. Some others, on occasion, would appear, literally by the millions, last a short while, then disappear forever into the recesses of time. But mostly the same old species there at the beginning were there at the end. And they were identical in most respects to the way they were when they first started.

But what was the "beginning" and the "end" of this eight-communi-tied, 200+ species of the Hamilton fauna (known from as far west as Iowa, and from Canada south along the Appalachians at least as far as Virginia)? For this is no isolated story; it is a true, repeated *pattern*, the most compelling and at the same time underappreciated pattern in the annals of biological evolutionary history. Brett and Baird have documented no fewer than eight of these faunas, appearing in succession in the Paleozoic rocks of the Appalachian region. Each one lasts some 5- to 7-million years. Each one formed from the seeds of just 20 percent of the species preceding them (their beginning), and disappeared abruptly as physical environmental factors overwhelmed the ecosystem, driving most component species to extinction (their end). Species coming in from elsewhere very commonly provided at least some of the new elements of the next succeeding fauna.

Again, this is not isolated phenomenology, this is repeated pattern. Throughout the 540-million-year history of complex life, we see again and again almost monotonous ecosystem stability. This monotony is interrupted only occasionally, but inevitably (in the long run), by ecosystem-generalized punctuations of environmental disturbance, followed by extinction, and, finally, speciation. And like George Simpson before them (and James Hutton before *him!*), paleobiologists wrestling with these patterns of systemwide stability and abrupt change have developed new theory. They have dealt *not* with newly imagined processes wholly undreamt of in traditional biological discourse, but rather with novel combinations of otherwise well-known phenomena that might account for the patterns they see. Patterns like the Hamilton fauna, ensconced within a series of eight such stable interludes that begin and end with a geologically sudden shock from the world of matter-in-motion.

Brett and Baird have introduced the term *coordinated stasis* for the sequence of Hamilton-like faunas in the Paleozoic of the Appalachian Basin.[15] The term itself, as some commentators have noted, implies an ecological control over stasis. It suggests that perhaps the internal "locking" of ecosystems, the long-term result of the putative highly interdependent nature of the relationship between components of ecosystems, is the ultimate cause of stasis within individual lineages. Others, such as myself, prefer explanations that stress habitat tracking and the Sewall Wrightian notion of internal species' structure that conjoin to produce stasis within individual lineages. Coordinated stasis as a pattern, in this sense, simply means that many different species' lineages simultaneously are in stasis. We have, in other words, a fascinating chicken-or-egg debate. One side sees ultimate control of the empirical pattern as the result of ecological processes; the other sees it as the result of internal species' structure and its relation to the physical and biological environment. But everyone agrees that the patterns are real, are to be found everywhere up and down the stratigraphic column (that is, over the entire history of life), and are framed, actually triggered, by events in the external, physical world. These physical events were too intense, in strength or duration, to be absorbed by the resiliency of the ecological-evolutionary system.

Yale paleobiologist Elisabeth S. Vrba has documented precisely the same sets of evolutionary-ecological sequences in the Pliocene of eastern and southern Africa. She has clearly implicated human evolution, patterns of extinction and evolution in hominid phylogeny, along with all the rest of the biota, in an event that began as global temperatures dropped some 10 to 15 degrees Centigrade over a period of some 200,000 to 300,000 years, beginning roughly 2.7 or 2.8 million years ago.

Vrba's term for the pattern is *turnover pulse*. Her explanatory hypothesis is a model of original combination of otherwise familiar ecological and evolutionary processes. In short, Vrba suggests that sudden disappearances, followed by equally dramatic appearances of new species, have two symmetrical sets of causes. One is simple habitat tracking: 2.5 million years ago, the habitats of eastern and southern Africa were characteristically moist woodlands. The wet woodlands absorbed the cooling, and associated drying, as far as they could. Then, all of a sudden, the woodland plants no longer could thrive; they were quickly replaced by savannah grasses with isolated patches of woodlands. With the woodland plant species went the vertebrate and other animal species adapted to them. Generalists, like impalas, survived the transition. And with the grasslands, in came savannah-adapted pigs, antelopes, carnivores, and so forth. In a similar fashion, cheetahs right now are beginning to appear in some numbers in Botswana's Okavango Delta system because the delta is drying up. The cheetah, for the first time at least in the memory of the delta's human inhabitants, are finding the wide-open savannah-type parkland they require.

But habitat tracking is only half of Vrba's story. Habitat tracking explains why evolution *isn't* occurring. The second component of Vrba's turnover pulse hypothesis suggests why evolution *does* occur under just these circumstances. First, extinction does take place: Not all species disappear from a region because they can find recognizable habitat elsewhere. Some of them just become extinct. And here is the Huttonian part of Vrba's idea that I like so much: If habitat disruption can lead to extinction, it can also lead to speciation. The rhetoric I grew up with called for evolution to "fill empty niches." But I never understood how that would work, if speciation, that is, is the matter of splitting one reproductive community, a species, into two. *Successful* speciation, the survival of fledgling species, does seem to be correlated with successful occupation of niches sufficiently different from parental species to allow the fledgling to survive. What Vrba is saying, in addition, is that habitat change—"degradation," from the standpoint of what came before—can lead to extinction, *but it can also lead to speciation.* Habitat fragmentation, particularly relatively abrupt, physically induced fragmentation, is just what is needed to cause reproductive fragmentation—in other words, speciation. In other words, evolution.

Nor is this pattern of general turnover purely paleontological. Looking at it strictly from the standpoint of the distributions of species in the modern tropical world, biologists have long thought that Pleistocene tropical habitats were disrupted by climate change. The *refuge hypothesis*, as developed especially by ornithologist Jurgen Hafner for patterns of bird species distribution in the Amazon Basin, postulates the spread of savannahlike grasslands during times of glacial maxima. These are precisely the same conditions that Vrba and other paleontologists documented from Africa 2.5 million years ago. Here, the evidence is disjunct distributions of closely related species of forest birds and other taxa. The savannahs have been invoked as ecological disrupters of once- (and now-) continuous tropical forest, thus neatly accounting for the disjunct distributions of closely related species. Hafner's refuge hypothesis is the striking mirror image of Vrba's turnover pulse.[16]

This general pattern—call it what you will: coordinated stasis, turnover pulse, or refuge—is the key to understanding the connection between the biotic world and the physical world of matter-in-motion of which life is a part. Linked, on the lower scale, to patterns of physiological resiliency of the organism, ecological succession, and habitat tracking, and on the upper to patterns of mass extinction and evolutionary response, this coordinated stasis pattern is the single most important ingredient pointing the way to a theoretical understanding of just what that connection is.

From Concrete Pattern to Abstract Theory: Evolution and the Realm of Matter-in-Motion

Once again, pattern in nature literally forces us to formulate new theory, abstract descriptions of the nature of systems and their characteristic behaviors. We have before us patterns galore crying out for linkage between the traditional realm of evolutionary explanation and the "physical" world. What is the underlying theoretical structure that simultaneously generalizes and specifies the actual nature of that connection?

It is clear almost at a glance that the neo-Darwinian paradigm alone cannot and will not suffice. Despite recent setbacks at the hands of ultra-Darwinians, that fact was made clear 50 years ago when Dobzhansky and Mayr established the importance of discontinuity between species as a real evolutionary phenomenon and not, for example, an artifact simply of extinction of intervening species. All the patterns of the preceding section force us to see species as discrete entities in space and time, ones with beginnings, histories, and ends. It was the patterns underlying the notion of punctuated equilibria, patterns of within-species stability and among-species evolutionary change, that transformed the Dobzhansky-Mayr notions of species into a full-blown historical concept: Species are spatiotemporally bounded entities.

But, as remarked at the end of the preceding chapter, punctuated equilibria was devised to cover patterns of within- and among-species stasis and change *within monophyletic lineages*. And all the patterns adduced in the preceding section involve more than one species, the vast majority of which are not closely related, living together in associations ("ecosystems") in specific geographic regions. These patterns, though they may have their phylogenetically coherent components (for example, the latitudinal diversity gradient works for most ecosystems *and* for most taxa), are patently, expressly cross-genealogical in character.

Thus Dobzhansky's powerful concept of the evolutionary hierarchy, what I and others have in recent years have called the genealogical hierarchy, does not in itself provide the links to the world of matter-in-motion. Some additional structure, one that explicitly accounts for cross-genealogical patterns in evolutionary history, needs to be elucidated before those connections truly can be said to be specified.

That additional, missing element is the *ecological hierarchy*.[17] Organisms do two distinct sorts of things: they live and they reproduce. In living, they extract energy and nutrients, storing that energy and utilizing it to construct proteins and other structural compounds in the development, growth, and maintenance of the soma.

"More-making" (as we saw in the last chapter) is the hallmark of the genealogical hierarchy: genes replicate, organisms reproduce, species speciate. Entities at any one level of the genealogical hierarchy go on making more entities of like kind. They are the components of the next-higher level: Genes are parts of organisms; organisms parts of breeding populations (demes); demes parts of species; species parts of higher taxa. It is the ongoing "more-making" of entities at any particular level that create the elements—give shape to, and keep "alive"—those of the next-higher level. Without speciation there would be no evolutionary strings of species, no higher taxa. If speciation stops, extinction eventually claims all members of a higher taxon. And that higher taxon, be it a family, an order, even a phylum, will go extinct. A similar thing occurs at the lower level: no organismic reproduction, no reproductive community; no genic replication, no organismic reproduction.

The ecological hierarchy is also a parts-whole structure. The soma, the nonreproductive "body" of organisms, consists of from one to billions of cells, often highly differentiated. It is the cells of the body that perform the basic chemical steps of energy transfer, storage, and material construction of the components of the cells, tissues, and organs. Indeed, as the preceding phrase suggests, complex, multicellular organisms are themselves hierarchical arrays of differentiated cells gathered together to form coherent, functional tissues, which themselves are gathered together to form specialized organs, which then form organ systems. The whole works together to produce a single, integrated whole: an organism.

But of prime interest here is the energy-acquiring activities of organisms, their "economic" interests and activities. Plants photosynthesize, literally "eating" sunlight. Except as noted (note 12, this chapter), photosynthesis is the source of the energy that flows among organisms within ecosystems. The sheer business of staying alive and making a living, procuring nutrients and food (that is, energy resources), and avoiding predation (including infection by pathogens) sets off a chain of interactions that leads to the existence of structured ecosystems. These systems are then linked through interactive energy flow regionally to form progressively larger-scale systems.

First off, organisms will interact economically with others of the same species. They variously cooperate or compete for energy resources. It is this economically interactive group that plays a concerted role in the dynamic flow of matter and energy in a local ecosystem. Such local populations of organisms of the same species in this explicitly economic sense are aptly termed *avatars*, meaning local representatives of a particular species. Each avatar plays a specifiable role, thus each has its own "niche" within that local ecosystem.

Avatars and demes are both local populations of a specific species. Clearly, the deme, the local breeding population, may be virtually the same as a local avatar. But there are real distinctions. Primarily, of course, they differ in their functions, but they also have subtle differences in composition. Two examples will serve to clarify: mussels and mammalian bulls.

Consider rooted colonies of mussels lining a rocky promontory along the northern California shoreline. Each and every individual mussel will spend the part of the day when it is submerged extracting oxygen, nutrients, and minute particles of food from the ambient seawater; each will be emitting carbon dioxide and other waste products; each will withdraw into its protective shell when the tide drops—and each will do so to avoid predation that could come from the sea (in the form of carnivorous snails, crabs, starfish, and others) or from land (in the form of gulls and other seabirds and humans). Each mussel is actively living an economic life every moment it sits there and stays alive in its little allotted corner of rocky intertidal shoreline. And the aggregate of those mussels, the avatar, is changing the distribution of ambient chemicals, nutrients, and food particles, and is, as well, serving as an energy resource for a variety of other avatars.

Yet when the reproductive season comes, some mussels will not be mature, others past their reproductive prime, and still others incapable of reproduction. The local deme of mussels on that rocky promontory is not exactly the same as the local economic avatar that is part of that intertidal ecosystem.

In some other instances, the lack of identity between the local economic avatar and the reproductive deme of same-species organisms is even more striking. In many mammalian groups, bulls of reproductive age stay well away from the main herds of females with their young, coming around only during the mating season. In African buffaloes, there is even the further distinction of expulsion of past-their-prime bulls from the main herd where once they were dominant. These expelled bulls tend to band together in small herds of four or five and are known as the meanest creatures on the African savannah. But whether it be lone bull elephants or small clusters of no-longer-breeding buffalo bulls, the ecological effect is generically the same: The avatar structure of African elephants and buffaloes for the majority of the year is very different than the structure of the actual deme during reproductive season.

It is at the next-higher level that the economic and genealogical hierarchies draw completely apart. If demes are splitting and merging as parts of the largest-scale reproductive community—species, the next level up in the genealogical hierarchy—avatars of one species are interacting with avatars of many other species to form the biological components of local ecosystems. And there is a world of difference between species, reproductive communities, essentially large-scale "packages" of genetic information, and ecosystems, composed as they are biologically of interacting avatars drawn from many different species.

If genetic information flow and the ongoing reproductive activities of component demes and organisms is what keeps a species up and running, it is the moment-by-moment physical, economic interactions between the microbial, fungal, plant, and animal avatars that makes an ecosystem go. Vital, too, to the internal dynamics of ecosystems are the purely physical aspects of climate (itself often regulated by ecosystems regionally) and, of course, the ultimate physical source of energy. In short, ecosystems constitute a world of matter-energy transfer, and a world utterly different from that other collectivity of populations: species. As spelled out in Chapter 5, species are not parts of economic systems; species don't have "niches." Rather, species are storehouses of genetic information, albeit much of that information pertaining to the environment.

Ecologists sometimes worry about the "reality" of ecosystems. Are ecosystems, in other words, spatiotemporally discrete entities as species appear to be? For though the edge of a stream might provide a sharp boundary between its rushing waters and the woodland on either bank, other boundaries—between dense forests, say, and open grasslands—might definitely be more gradational. Are such so-called ecotones to be considered part of one or both of the ecosystems? Or do they constitute their own quasidiscrete system?

Compounding the problem of specifying the boundaries of local ecosystems is the patchiness that comes from the ongoing process of degradation. Periodically, patches of landscape are disturbed. For example, woodlands may be disturbed by fire or windstorms. In the longer term, generally less-localized areas may be affected by flooding, drought, or even climate change, including glaciation. In more localized instances, the ecological clock of succession is reset, as when alders and other pioneer species establish the first foothold, followed in regular order by other plant avatars, leading ultimately to the restoration of "mature" forest. Thus a patch of alder within an otherwise mature evergreen forest in the Rocky Mountains may be a sign of disturbance followed by the earliest stages of successional recovery. These may be aspects of the same, rather than a sign of two distinct, ecosystems.

Boundary recognition problems aside, there is no doubt that localized systems do exist where matter-energy transfer between avatars is concentrated. And they do so with less such interactive contact occurring with avatars within other, adjacent local systems. However, virtually no ecosystem is an island, cut off from all the rest. Hawks perching on woodland trees nab mice scampering across grassland floors. Energy flows across the often vaguely defined boundaries of local ecosystems, linking up local ecosystems into an ever-widening network of regional, then multiregional, and eventually continental and global dimensions. If species are parts of higher taxa, and if all higher taxa are linked up into a hierarchical array of more-inclusive groups (Kingdom Plantae, Kingdom Animalia, and so on), eventually to fall under the single, though not officially named, taxon "All Life," so too are ecosystems. *Gaia* is the most commonly encountered name for this global ecosystem.[18]

So there we have two distinct separate, hierarchically arranged organizational systems of living things, as shown in the figure below. Each flows from one of the two basic activities of organisms: reproduction, yielding the genealogical hierarchy; and economics, the business of ob-

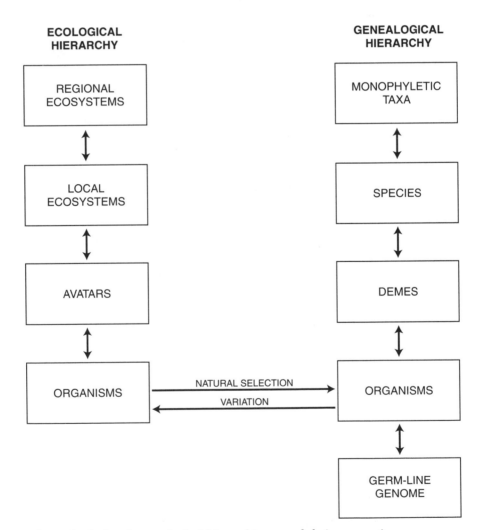

The ecological and genealogical hierarchies — and their connection.

taining energy and dealing with life's other exigencies, yielding the "eco-logic" (or, alternatively, the "economic") hierarchy. The former, adumbrated by Dobzhansky, is clearly concerned with the transmission and storage of genetic information. The latter just as clearly is connected with the physical world, the world of matter-energy transfer.[19] Connecting these two hierarchies will show us how the domain of evolution connects to the world of matter-in-motion. It will yield the theoretical structure we need to fold the kinds of patterns reviewed in the preceding section coherently into evolutionary theory.

Connecting the Genealogical and Ecological Hierarchies

Organisms, or course, are parts of both hierarchies; it is their activities that lie at the hearts of each of the two systems. Yet ever since Weismann, with his distinction between soma and germ-line, organisms themselves seem to be alliances (at times even uneasily so) between the economic business part of their lives and bodies, on the one hand, and the reproductive side on the other. Nor is the alliance symmetrical. Notoriously, reproduction is the only category in the classic list of physiological functions such as respiration, excretion, and digestion absolutely not necessary to the survival of the individual organism. On the other hand, reproduction requires an energy-procuring and dispensing soma.

Organisms don't evolve. They develop, live, and die. But that is not evolution. It is not evolution because, despite the occurrence of both germ-line and somatic mutations, and no matter how greatly these (as well as environmental and simple aging factors) may change the appearance of an organism over the course of its lifetime, the genetic information fades as death degrades everything. Certainly the individual's DNA and RNA are included in the degradation. Evolution may well be defined as the proposition that all organisms are descended from a single common ancestor, "descent with modification." Nevertheless in modern parlance, evolution amounts to the fate of genetic information. When an organism dies, its genetic information has one of two possible fates: either it dies with the organism (that is, has a *genetic death*), or it survives in some subset in descendant organisms. It's not what happens to genetic information *within an organism* (except for germ-line mutations), it's what happens to that information *among organisms* that is grist for the evolutionary mill: Is it passed along to descendants? Or does the organism die without reproducing?

Natural selection is clearly the main connection between the ecological and genealogical hierarchical systems. Consider the figure on page 166. The reproductive behavior of organisms within demes continually produces new organisms, with a spectrum of expressed, genetically based anatomies, physiologies, and behaviors. These new organisms join their elders, staffing the local avatars (which may or may not be nearly identical with the local deme). Purely in terms of the economic adaptations of organisms, which make up the vast bulk of any organism's adaptations, some of these new recruits fare better than others in the economic game of life. Purely as a side effect, the differential economic success among peers within an avatar shows up, statistically, on the reproductive success these organisms have back on the genealogical, demal side of the ledger.

And it is indeed a ledger. The foregoing synopsis is simply a rendering of classical Darwinian natural selection. The genealogical hierarchy sequesters genetic information. Ongoing reproduction of organisms within demes constantly presents new recruits to the actual arena where the game of life is played: local ecosystems. Species themselves are merely larger-scale banks of genetic information. When avatars are wiped out, as portions of regional ecosystems are degraded (after, say, a fire or an oil spill, or also under the due effects of normal ecological succession), it is the demal side of a species' population structure that supplies the raw recruits when the area is recolonized. It is the demal aspects of species population structure, moreover, that sends the "propagules" out as the map of species distributions changes, reflecting patterns of climatic change.

Evolution for its greatest part is the fate of genetic information in an economic context. But it is more than natural selection (and genetic drift) in the context of demes and avatars. It has to be, for species are also packages of genetic information, and as the preceding patterns make clear, evolution also entails differential births and deaths of entire species. Interestingly, unlike the near-identity of demes and avatars, which involve two functionally different aspects of the population structure of species, the connections between species and the economic realm are not clear.

Indeed, I have taken great pains to show the distinctness of species and ecological entities. For one thing, species are not parts of ecosystems. Rather, local ecosystems and species are the next step up above avatars and demes. The former consists of moment-by-moment economic interactivity of avatars, the other the occasional but ongoing reproductive activities of organisms organized into demes. Both ecosystems and demes, in other words, consist of interacting populations. But the nature of the populations and their interactions are radically different. So, we ask, how *does* the economic world impinge on the fates of species?

Part of the answer is straightforward and conventionally reductive: The genetic composition of species changes over time as a net effect of generation-by-generation selection and drift-mediated, mutationally based (ultimately) genetic variation. But species, as the overwhelming pattern, do not change all that much, if demonstrably at all, in terms of their preserved economic adaptations (anatomies). The combined avatar-mediated changes in demal genetics (with avatars present in many different ecosystems) typically sums to zero change. That, and the propensity of demes to send out "seedlings" to newly habitable regions, is apparently sufficient to keep species stable even in the face of fairly gross environmental change.

But species do speciate, and they do go extinct. And, if the patterns recounted above are any indication, as they surely must be, species do so typically in lockstep. And they do so not so much with other species of their own lineage but rather with other species whose avatars share ecological space with them.

Moving still farther up the scale, we start noting a symmetry between degrees of effects on the ecological side with their effects on the genealogical side, and vice versa. It is as if the connections between the ecological and evolutionary hierarchies are a "sloshing bucket": The higher up the slopes of one hierarchy, the higher up the resultant effect is likely to be on the other.

Regional ecosystemic degradation such as Vrba's 2.5 million-year-ago African event typically have higher stakes: Species drop out, some through emigration, some to true extinction. As the system reassembles, other species come in, from immigration, and from actual speciation. There is symmetry, in other words, to the degree of loss and the degree to which systems must go to assume a semblance of normalcy.

That symmetry is met at the highest degree during global mass extinctions. For example, when higher taxa themselves disappear, only to be replaced eventually by ecological-role-playing equivalents, systematists invariably assign equal, or nearly equal, rank to the replacement taxa. Witness, for example, the demise of the order Rugosa, the Paleozoic corals that became extinct in the Permo-Triassic global extinction event. They had been both the solitary and the reef-building corals of the Paleozoic. After the usual hiatus, modern scleractinian corals first appeared in the Lower Triassic. They were derivatives, apparently, of the stock of naked anthozoan coelenterates that were probably always there during the Paleozoic, and which still grace the world's shorelines: sea anemones. Both Rugosa and Scleractinia are considered orders of the class Anthozoa, phylum Coelenterata.

Thus, as I read this diagram, the more encompassing the ecosystemic shock, the farther up the opposite, genealogical hierarchy the shockwaves travel. Only demes are eliminated in local events; whole orders, classes, possibly even phyla are eliminated in more severe and widespread events. Consequently, in "bucket-sloshing" mirror imagery, the more different will be the possible candidates on which speciation and natural selection work to shape the new ecological systems. If the old corals are gone, new ones take their place, drawn from remote collateral kin. Had the anthozoans disappeared completely at the end of the Paleozoic, possibly the hydrozoans (a separate class of coelenterates) could have filled the bill. Possibly not. Possibly the sponges would have been left to provide the building blocks for the massive calcareous nearshore organic structures we call "reefs," which have been present for most of the past half-billion years, built by waves of a successive cast of characters. Or possibly the Permo-Triassic extinction could have been even more encompassing, leaving literally no viable candidates in the role of block-producing reef builders. In which case the oceans today, and the evolution of many groups, advanced bony fishes among them, would have been utterly different than the (highly threatened) diversity we have around us today.

The stronger the shock, the more devastating the ecosystemic degradation. The more devastating the degradation, the farther up the taxonomic ladder the removal of packages of genetic information. The higher that level of removal, the more different the remaining genetic information to reshape the diversity of life once the extinction vector is dampened. And the greater the disparity in genetic information from what had once shaped the ecosystems of the world, the more different the new components, hence the new ecosystems, will be. That's the sloshing-bucket model of interactions between the two hierarchies. That to me, in a nutshell, is how life evolves.[20]

We have come a long way down the *Cecropia* trail. I have by now answered my initial questions at the trail's very beginning. And we have seen that there are two radically different answers to the main question: What drives evolution? To me, a somewhat typical paleobiologist of the late twentieth century, life pretty much stays stable unless something comes along to stir things up, to "slosh the water in the bucket." That something, typically, is physical, like meteors slamming into the earth, or great volumes of volcanic ejecta, or the more subtle but no less ultimately devastating effects of global climate change. (Though today the equivalent vector of extinction is the devastation caused by a single species, *Homo sapiens*, because that species no longer lives as avatars within local ecosystems.) Then, and only then, will life reassemble the evolutionary differences reflecting the degree to which new avatars are recruited from different phylogenetic stocks.

But there is still one more stop along the *Cecropia* trail. For we can do more to link up, in a systematic fashion, our understanding of sources of energy on earth with the sloshing bucket of ecological and evolutionary resonance.

Plate Tectonics, Ecology, and Evolution

As recently as some 5 million years ago the Mediterranean Sea dried up completely, an astonishing event revealed by several sets of patterns that at first glance appeared unrelated. Seismic profiles of Mediterranean bottom sediments in the 1960s persistently turned up an apparently continuous hard, highly reflective layer that turned out to be salt. This was prima facie evidence that the Mediterranean, over 5 miles deep in places, had been reduced to a gigantic salt pan. A bit earlier, Russian engineers drilling the footings for the Aswan High Dam some 600 miles up the Nile from the Mediterranean found the Nile Valley to be much deeper than predicted. This discovery implied that the valley eroded down to a former sea level very much lower than today's level. There is no doubt that the Mediterranean dried up 5 million years ago.

Nor is there any doubt as to the cause of this event: a relatively minor movement of the tectonic plates at Gibraltar closing the connection between the Atlantic and the Mediterranean. This movement cut off the supply of "fresh" oceanic waters that constantly stream into the Mediterranean. Even today, evaporation is so high, and the freshwater riverine runoff too insufficient to replenish it, so that the eastern Mediterranean is very salty. The Mediterranean would very quickly dry up again should Gibraltar once more close.[21]

The biotic consequences of this event were significant: death to all marine creatures, including extinction for indigenous (endemic) species. An evolutionary radiation of cockles and other mollusks also occurred in the newly isolated Pontian Basin; and antelopes, tracking the spread of more open-ranged habitat, reached Africa from Asia. (A cool, dry period, not known to be associated with actual glaciation, coincided with the drying of the Mediterranean.)

A much clearer picture emerges with the next phase of global cooling. Plate tectonics are linked with global climate change, and hence with the disturbance of ecosystems and the distribution, extinction, and evolution of plant and animal species. This phase is the aforementioned 2.8- to 2.5-million-year cooling that triggered Elisabeth Vrba's "turnover pulse" in the African biota around that time. The story begins with the closure of the Isthmus of Panama, the culmination of 15 million years of geotectonic development that transformed the region from a deep, open oceanic basin into a volcanic arc at the margin of the Pacific and Caribbean plates, through accretion of exotic terranes (crustal blocks), and eventual collision with South America. The Isthmus became a fully connected terrestrial barrier between the Pacific and Caribbean somewhere between 3.1 and 2.8 million years ago.[22]

The biotic consequences of this closure are long known and well appreciated, at least to some extent. Most obviously, land biota was able to mix much more freely, and the influx of alien species caused great ecosystem disruption and the extinction of many endemic species. (Most famously, North American placental mammals invaded South America, though the migratory vector ran the other way too.) Somewhat less clear was the effect of isolation on the marine species of the Caribbean and Pacific. Jeremy Jackson and Ann Budd (see note 23, this chapter) dispute some of the prior claims that *geminate* marine species, closely related "sister species," soon appeared on either side of the isthmus. But they do also note that "most biological change occurs in pulses." Specifically: "The marine biota of tropical and subtropical America was dramatically transformed roughly 2 million years ago, although the tempo, and perhaps the timing, of the turnover varied within and among taxa."[23]

Jackson and Budd trace the dramatic transformation of the marine biota to intensification of northern-hemisphere glaciation, the very cause of Vrba's turnover pulse. And here is where the causal nexus becomes very intriguing indeed. Though the actual causes of glaciation are still not completely understood, climatologists and paleo-oceanographers have begun to suspect that it was the closure of the Isthmus of Panama that did it, for closure meant the deflection of old patterns of oceanic circulation. In particular, closure meant the development of the Gulf Stream, bringing warm, saline tropical waters to the higher latitudes, perhaps triggering the buildup of ice around 2.5 million years ago.

Consider other examples: As we saw in Chapter 3, the patterns of both evolutionary relationships *and* geographic distribution of mockingbirds on the various Galapagos Islands are a direct reflection of the relative ages of the islands, the oldest in the southeast, the youngest in the northwest, as the region has drifted over a stationary "hot spot," a mantle plume. The Hawaiian Islands have formed much the same way, with evolutionary and bio-geographic patterns of its famed flora and fauna also reflecting the geotectonic history.

A final example: Lake Victoria has just recently been discovered to have dried up completely only some 12,000 years ago, the result of tectonic shifting of its drainage patterns. Thus at least most of its more than 200 species of endemic cichlid fishes must have evolved in a rapid flurry of speciation once the waters returned.

What reads like a litany of case after case of isolated examples merges into a steady stream of repeated pattern: Regional tectonic changes in the earth's crust modify patterns of oceanic circulation, having profound effects on the atmosphere and hence on global climate.[24] Distributions of species are directly affected, ecosystems are modified, and ultimately extinction and later speciation follow. The patterns known to our nineteenth- and early twentieth-century predecessors — patterns that ultimately led them to accept the basic notion that life evolved, plus historical biological patterns that contributed to the eventual acceptance of continental drift and plate tectonics are telling us now that the basic control of the evolution of life on earth *is* plate tectonics and its subsequent effects on oceanic circulation and global climate.

And here we meet up, once again, with the remarkable James Hutton. For the linkage between the physical realm and life and its evolutionary history at first glance seems as though it must come through the matter-energy transfer, physiological behaviors of organisms. Therefore, when that link was sought earlier in this chapter, it was natural to have started with latitudinal diversity gradients, which have held out the promise of connections between biological patterns of distribution and evolution with patterns of distribution of solar energy over the surface of the earth.

And such connections are real. But deeper still, literally, is the connection between *endogenous* energy, heat flow from deep within the earth, and biological evolution. It is endogenous energy that drives the plates, which shapes the surface of the land and the conformation of the oceanic depths. It is the changes in that physical landscape, with all their myriad consequences in nutrient distribution, rainfall, and ambient temperature, that ultimately control events in the evolution of life. Nor is this anything new in the history of life. If the events cited by paleobiologist Joe Kirschvink (see Chapter 4) are any guide, physical event-driven evolution has apparently been the norm since life's inception. Almost unimaginably extreme events have occurred, such as "snowball earth" (glaciation reaching the tropics) and the even wilder shift of the entire crust by 90 degrees, each apparently correlated with a major event in the evolution of life.

This is not to say that no evolution of any kind can take place unless triggered by climatic and other tectonically induced causes. For one thing, the invasion of entirely new kinds of environments is at least a step removed from the more direct evolutionary response to physical events: vascular plants could only colonize terrestrial realms after advanced algae, chlorophytes, had themselves evolved. And it was only after plants had become established that insects then could evolve. In such circumstances, the potentially invadable habitat was long in existence prior to the evolution of the first invaders. And those invaders, plants in this case, had to modify the environment and provide a basis of existence for themselves before the other major components of the terrestrial biome could evolve.

Nor am I suggesting that biologists focusing on the neo-Darwinian paradigm of adaptation through selection in local populations are somehow misguided, or even that the metaphor of the "selfish gene" has no place in evolutionary discourse. The mechanics of genetic drift and selection-mediated change of gene frequencies in populations remain as central as ever to evolutionary thinking today.

But what I am saying in no uncertain terms is that simple extrapolation upwards of a competitive model for transmission of genetic information from one generation to the next does not suffice to explain more than a fraction of the commonly encountered patterns in the evolutionary history of life. The fault in this reductionist, extrapolationist vision lies in the penchant for "arm-waving," for seeing physical events as isolated incidents that set selection off in this direction or that. This tired old "in principle" explanation of how the evolutionary process works does not do the job. It does not explicitly link, in a general theoretical framework, evolutionary biological systems with the rest of the physical world. Indeed, to insist upon a form of internalized competition among genes themselves for representation in the next generation as the basic motor of evolutionary change is to substitute a pseudo-physical model, while abandoning the search for law-like links to the actual physical universe. As we have seen, evolutionary biology remains wedded to a curiously internalized vision: Competition for genetic representation in the next generation is said to arise among the genes themselves. And even when the "environment" is explicitly considered, it is most frequently the presence of other organisms—members of the same species, or of other avatars—that are cited.

But the physical world is more than simple backdrop to the evolution of life. The physical world changes in a regular, intelligible manner. And those changes have had profound effects on the evolution of life. These evolutionary effects are regular and lawlike. And they invite—actually demand—rational contemplation and explicit incorporation into evolutionary theory.

Adding the hierarchy of ecology to an expanded version of Dobzhansky's hierarchy of genealogy helps to expand evolutionary theory, pointing it away from a purely lineage-based discourse towards the rest of the physical world. But now that plate tectonics has melded the study of the patterns of geological history with an understanding of the dynamic processes that produce those patterns; and now that progress has also been made in integrating, in good Simpsonian fashion, patterns of biological history with theoretical understanding of the evolutionary process, the time has come to acknowledge and embrace the physical context of biological systems—and the evolutionary history of life.

Notes

Chapter 1

1. The passage in question occurs on page 90 of Dawkins's *The Selfish Gene* (1976). There he claims that the structure of ecosystems follows from simple competition for reproductive success among organisms—or for replicative success among genes. I have characterized and criticized as part of my analysis of his views the virtues and failings of what I call the "ultra-Darwinian" notions of Dawkins and other biologists. For details, see especially my *Unfinished Synthesis* (1985a), *Macroevolutionary Patterns and Evolutionary Dynamics* (1989) and most recently *Reinventing Darwin* (1995a), where I explore details of the history of the debates between ultra-Darwinians and paleontologists such as myself.

2. For an excellent discussion of the biology of *Cecropia* trees, and of neotropical ecosystems generally, see John C. Kricher's, *A Neotropical Companion* (1989), especially page 85 and beyond.

3. I use the term *you* to mean any organisms of any species, microbe, plant, fungus, or animal, but certainly including individual humans.

4. I thank my colleague Margaret Wertheim for this term for the physical world, the realm studied by physical science. Such a world, I hold throughout this narrative, has not yet been fully integrated with evolutionary biology.

5. See Bunge (1977).

6. Mayr contrasts *historical* with what he terms *functional* science especially in Mayr (1961) and in his monumental work on biological history, *The Growth of Biological Thought* (1982). The latter work is the first of a projected two-part study of the history of biology divided into historical biology (the volume published in 1982) and functional biology (said, in the introduction to the 1982 volume, to be forthcoming). For Simpson's views on the supposed distinction between historical and "nonhistorical" science, see especially Simpson (1963). I have discussed their views, and my own take on the subject, in depth in Eldredge (1993).

7. The famous offending remark, which ultimately sparked such good work, is in Simpson, 1961, p. 130.

8. See Platnick and Cameron, 1977, for an excellent, very revealing discussion of the isomorphism between the principles of stemmatics, historical linguistics, and

phylogenetic systematics. Realization of the similarities among these disparate fields arose when H. Don Cameron, a historical linguist at the University of Michigan, discovered the writings of Norman Platnick, an expert on spider systematics at The American Museum of Natural History, and an early exponent of the then-new cladistics. Cameron, it seems, was also a spider aficionado, volunteering his services to curate The University of Michigan Museum of Zoology's spider collection.

9. The most notorious case, possibly apocryphal, of such errors introduced in manuscript copying is the apparent explanation of the story of Saint James's visit to Spain—where allegedly Hierusalem somehow became transmogrified into Hispania. The history of European pilgrimages to Santiago de Campostela is a not-so-trivial consequence of this typo. Mutations of messages are, of course, well-known phenomena, as in the old parlor game commonly known as Telephone. There, one person whispers a message in a neighbor's ear, with the injunction to "Pass it on!" Inevitably a completely different message emerges at the end after just a few such iterations.

Closer to home, a pioneering ultra-Darwinian biologist recounts a history of such errors. William Hamilton's work laid the very foundations for sociobiology and much of the gene-centered perspective of ultra-Darwinian biology in general—thus his work is often cited. He says in his recent book (Hamilton, 1996, p. 21) that "an evolution of errors" in the citation of two of his most important papers (Hamilton 1964a and 1964b) proves that a very high proportion of the citing was being done "by people who had not read the work." It probably is being done still.

Actually, reading a work is no guarantee of an accurate bibliographic citation, which is usually done much later, and from someone else's handy bibliography. For the record, my own citation herein of these particular papers of Hamilton is taken directly from his recent book. I read the original papers quite a few years ago. Nonetheless, I will not go to the original publications to verify the reference. Thus any error in my citations of these two historically important contributions belongs to either Hamilton, me, the editorial process at W. H. Freeman, or the printer.

10. In their pioneering work on biometrics, Simpson and Roe (1939; revised as Simpson, Roe, and Lewontin, 1960) discuss the tendency of many natural phenomena—for example, variation in biological structures, so critical to the workings of natural selection—to be for the most part "normally distributed." This alludes to a pattern of symmetrical variation around a mean value. They contend that, in the best of possible worlds, a sample size of at least several dozen is necessary to be confident of capturing the basic range in variation of, say, the dimensions of the first lower molar in a species of fossil mammal. But they also go on to make the very important additional point that the greatest shift in knowledge actually comes between a sample size of 0 and 1. It comes between utter ignorance of the existence of an entity, and a single example of it. It can thus be assumed to be drawn in the greatest likelihood from somewhere in the vicinity of the mean. Thus a pattern can be established by a single observation. Confidence, of course, grows through repeated observation.

11. For example, I (Eldredge, 1995b) have recently argued that the evolutionary history of hominids, a sequence of protohumans culminating in ourselves, records a progressive change from a biological adaptation-dominated mode of life to one mediated increasingly by culture. I have also claimed that, with the advent of the first agricultural revolution beginning some 10,000 years ago, in several scattered places independently, Homo sapiens became the first species no longer to live as broken-up subpopulations integrated into local ecosystems. But absolutely every other species on

earth continues to do so, and presumably has done so in the 3.5 billion years of life on earth. This factor thus removes the Malthusian cap on human population, enabling a present population of nearly 6 billion, a growth from roughly 5 million in just 10,000 years. I then argue that, with this development, we have become the first species to become a united economic force, interacting with the global energy-transfer system (that is, the global ecosystem, the sum of all local ecosystems). The only way such claims can be made is by noting the properties of species in general. In this case the fact is noted that all other species are broken up into subpopulations integrated into local ecosystems, and that it is such local populations, not species themselves, that play concerted economic roles (have "niches") in local ecosystems. These generalizations arise from systematic consideration of biological systems in general, and in particular of large-scale systems, especially in the context of hierarchy theory. That subject will be developed further in this chapter and in greater detail as a major theme throughout this narrative.

12. See especially Ghiselin, 1987. The use of the word *law* in discussions of what science is and is not has not been entirely salubrious. It has contributed to the inability to see the methodological link that pervades the entire enterprise, be the fields frankly "functional" and "ahistorical," or essentially "historical" in nature, in the sense that Mayr and Simpson had in mind. True laws such as gravitation are held to be immutable, at least in a Newtonian world, in contrast to the "laws" that apply to the messier world of biology, where exceptions seem to abound. For example, consider speciation, the fragmentation of reproductive communities into two "daughter" communities. It is for the most part consequent to a period of geographic isolation of parts of a single, ancestral species. This pattern permeates not just the modern biota but, in my view, the entire history of life. So much so that the statement "speciation requires geographic isolation" is a generalization with great scope and power. Exceptions to that generalization, such as the concept of *sympatric speciation*, are still under debate, and at most they are the exceptions that prove the rule.

13. There is, of course, a vast literature on perception. Closer to the points I raise here, however, is the work of N. R. Hanson (1958) and, more recently, the insights and discussions of philosopher Ronald H. Brady (Brady, 1994a, 1994b, and especially unpublished ms. [see Bibliography]). Brady's analysis of Goethe's thinking on the unity of mind and phenomena have been especially stimulating and helpful to me in formulating the views expressed here.

14. The distinction between the search for descriptions and explanations of agreed-upon patterns, on the one hand, and the development of new ways of seeing patterns on the other, as I develop it here, smacks of the distinction that Kuhn (1962) makes between "normal" science and the "revolutions" that lead to the development of new paradigms. I mean no such rigid distinction here, however. Important new discoveries well within an existing paradigm involve learning how to see patterns differently as often as they do coming up with better explanations of patterns commonly perceived. The following examples in the text, drawn from my own work on *punctuated equilibria*, which in my view fits wholly within the tradition of evolutionary biological discourse dating back to Darwin, I think amply demonstrate this.

15. See Eldredge (1971) and Eldredge and Gould (1972) for the earliest statements of punctuated equilibria. I give a much fuller account of the search for pattern and meaning in my research on the trilobite *Phacops rana*, work leading to the fundamental

elements of punctuated equilibria, in my book *Time Frames* (1985b; reprinted 1989). I continued my explicit arguments for the essential correctness of punctuated equilibria, and attendant implications, in my more recent *Reinventing Darwin* (1995a). My purpose here is very different: I seek merely to give examples of pattern perception and interpretation, as seen through some of my own personal experiences.

16. Trilobites are an extinct, early, and somewhat primitive group of marine arthropods that ranged throughout the entire Paleozoic Era, from its inception in the Early Cambrian Period (around 543 million years ago) until its end in the Late Permian Period, some 240 million years ago.

17. Principally, I subjected them to forms of factor analysis and multiple discriminant analysis.

18. See Eldredge and Grene, 1992. Grene's (1987) analysis of the nature and significance of hierarchies in biology, including her remark paraphrased here, is especially insightful.

19. See especially Simon, 1962.

20. Edmund Burke remarked to the effect that, though day fades gradually into night, on the whole the two are tolerably distinguishable. This observation is a favorite of mine and of those who take the messiness of boundaries, including intergradations between two systems, as posing no serious threat to the proposition that two large-scale entities do in fact exist.

21. See, for example, Nagel, 1961; see also Wimsatt, 1980.

22. See Eldredge, 1995a, and Lieberman, Brett, and Eldredge, 1995. The argument, in brief, is that natural selection takes place within populations integrated into local ecosystems. Each ecosystem being somewhat different, each local population is bound to have slightly different evolutionary histories. (This viewpoint was developed far earlier by Sewall Wright; see, for example, Wright, 1931, 1932.) Thus there is no way natural selection can "move" an entire species in any one particular evolutionary direction. For more on these issues, see Chapter 5.

For an extended discussion of upward and downward causation in hierarchical systems generally, including the notions of *initial* and *boundary conditions*, see especially Salthe, 1985.

Chapter 2

1. Just as massive parallel-processing computers are increasingly brought to bear to solve problems that require prodigious and otherwise prohibitive numbers of separate calculations (such as matching up gene sequences), the scientific community is increasingly called upon to tackle problems in which the sheer numbers of investigators are expected to prevail where their fewer, and less well-funded, predecessors so far had failed. The "war on cancer" is one such example of throwing more money and people at an admittedly important, if diffusely complex, problem that involves pure as well as applied scientific research. Increasingly, the scientific community sees itself in such light, as in the human genome project, or, even more recently, the call for scientific analysis and description of the world's inventory of species. The latter is urgently required, it is plausibly maintained, before the vast majority of species are driven to extinction in the current biodiversity crisis.

The importance of such tasks and utility of the massive labor-and-money approach in such circumstances notwithstanding, the essence of scientific creativity in the past has overwhelming lain with the individual. This remains true to this day and arguably for the foreseeable future. But team efforts are increasingly to be found at the heart of new discoveries and understandings, as the modern history of plate tectonics (see Chapter 4) and the more recent discovery of the top quark so plainly show.

2. In his *The Great Devonian Controversy* (1985), Martin Rudwick has painted an especially compelling and intriguing portrait of one such society, the Geological Society of London, still very much extant. Described are events from the 1830s and early 1840s in which the society's members wrangled among themselves, and with outsiders, over the classification of the Devonian System, the body of rocks corresponding to the Devonian Period of the geological timescale then under development. Through his meticulous analysis of a controversy that, once resolved, has faded from the collective memory of the geological profession, Rudwick has achieved marvelous insight into the workings of science in its critical early days before it became fully "professionalized."

3. These early days of concerted rational contemplation of the material realm, in one sense at least, seem deceptively simple. After all, almost any observation of a natural phenomenon was likely to contain novel elements. I was bemused some years ago to come across a paper from an early meeting of the American Association for the Advancement of Science in which a new species of trilobite was described. That the subject was brought up by the AAAS is testimony to the relative lack of discipline fragmentation back then. New trilobite species are still being described every year, but only in highly specialized journals. No paleontologist would expect to present a paper containing a description of a new species at a modern meeting of the AAAS, let alone submit a paper whose sole purpose was to describe it.

On the other hand, the early bird gets the worm—and early science was dedicated to the generation of not only then-novel minutiae but also some of the grandest of generalizations. The very notion of deep history of the earth and of life and of the evolutionary processes that have produced such histories, is one such grand generalization. Science writer John Horgan's (1996) thesis of "the end of science" takes up this concern, and it contains some substance and force: Grander themes and deeper truths are likely, though by no means guaranteed, to arise earlier, rather than later, in the history of rational contemplation of aspects of the material universe. In both geology and biology, no theme is likely to emerge again that is grander than the simple realization that the earth and life have had histories stretching back billions of years.

4. In one of his aborted attempts to find a profession, each prompted by his impatient father, Charles Darwin spent a year in Edinburgh as an entry-level medical student.

5. For details, see the 1956 Dover reprint of John Playfair's 1802 *Illustrations*, with informative prefatory commentary by G. W. White.

6. Darwin needed similar goading to publish a shorter version of his long-promised, but persistently unforthcoming, long work on organic evolution. His "abstract" was the respectably long *On the Origin of Species*. It was published, not because criticism

had prompted a clarification of his views, as was the case with Hutton, but because someone else, Alfred Russell Wallace, had independently come up with essentially the same theory of evolution through natural selection, in many instances right down to the same choice of terminology.

7. Gould was an undergraduate at Antioch College in Ohio. There his geological mentor was none other than George White, who wrote the introduction and biographical notes for the above-cited Dover edition of Playfair's *Illustrations*. The title of Gould's paper, "Is Uniformitarianism Necessary?" was a takeoff on James Thurber's "Is Sex Necessary?"

8. Noteworthy, too, was that Cuvier was one of very few members of the nobility to have his prestige as well as his head survive the French Revolution intact. His laboratories still stand on the northern edge of the Jardin des Plantes in Paris, where his legacy continues in the research efforts and public displays devoted to comparative anatomy, paleontology, geology, and—most recently—a gigantic display on the diversity of life on earth and the threat of a major extinction looming for the modern biota. This latter new exhibition, in the Grande Gallerie that had been closed to the public since the waning days of the nineteenth century, is especially fitting given the prominence Cuvier accorded extinction (his *Revolutions*) in the history of life. Its only rival is the Hall of Biodiversity, which opened on May 31, 1998, at the American Museum of Natural History in New York.

9. Though I address these issues in greater detail later, it is worth noting here that it is precisely this confusion over temporal scale that has fueled much of the debate between proponents of punctuated equilibria and those who defend a more orthodox Darwinian view of the evolutionary process. We punctuationalists have been routinely compared with *saltationists* (those who have proposed that evolution proceeds literally in overnight, single-mutational "jumps"). This, despite the fact that we punctuationalists have tried to make clear from the outset that we are talking about the much slower process of true speciation. Yet the debate over punctuated equilibria does not just involve a question of the rates of evolutionary change through natural selection. (For example, rates during the millennia of speciation were *relatively* fast versus the rates over subsequent millions of years, which were much slower.) Other, more qualitative issues are involved as well: for example, punctuationalists hold that the nature of large-scale entities, such as species themselves, must be taken into account before there can be a complete understanding of the nature of the evolutionary process.

10. Dobzhansky, 1937 (*Genetics and the Origin of Species*, first edition), page 12.

11. Literally as this passage was being written, the August 9, 1996, issue of *Science* carried the news story of a continuing battle among evolutionary geneticists. Some defend the orthodox position that competition for reproductive success among individual organisms alone suffices to account for evolutionary history. Others—a vocal, persistent minority—insist that selection acts on groups of organisms as well as on individuals.

12. A worse scenario occurs when the facts of history are denied. For example, both Cuvier's "revolutions" and the empirically well-established phenomenon of species stasis were abandoned in the wake of Darwin's gradualistic account of the evolutionary process.

13. And perhaps on Mars, and who knows where else, as recent stunning reports have it.

14. There are at least two and a half such accounts in Genesis. The most famous interpretation of creation events is Archbishop Ussher's 1654 analysis of all the "begats." His was a not-unclever way of using the genealogical account, plus ages, of the descendants of Adam and Eve to estimate exactly when God created the Earth. The calculation initially put creation at October 26, 4004 B.C. at 9 A.M. His account is the basis for the standard creationist estimate of the age of the earth: 8,000 to 10,000 years old. These creationists err, it seems, on the safe side by generously expanding the archbishop's original calculation.

In my experience, creationists care most about the biblical account of human origins. They often abandon a strict "young earth" position, and thereby sometimes concede a very old (and geologically correct) age to fossils of invertebrates and other organisms remote (they suppose) from human beings.

15. And, of course, they made their observations long before the conclusion that the earth was spherical became a generally accepted (western cultural) "fact." Incidentally, Eratosthenes calculated the circumference of the earth as the equivalent of 28,000 miles, roughly 4,000 miles more than the actual value, thus astonishingly close to present-day measurements. That the world is round was suggested by several commonly observed patterns. The Greeks, for example, observed that the earth cast a curved shadow on the moon and that mast tips are the first parts seen of any ship approaching port.

Knowing some of the details of celestial movement before realizing that our earth is round is a perfect example of the importance of scale in observation: The vast distances of space reduce enormous objects to manageable observable size. (Indeed, we need light and radio telescopes to blow them back up again, to make them truly visible.) Standing on the earth, so enormous to a single human, obscures one of its most basic features: its roundness.

16. See the second edition, 1965, pp. 159–160, for a discussion of Siccar Point and for all quotes.

17. Charles Lyell was something of a hero and mentor to the younger Charles Darwin. Much has been made of Darwin's disappointment that neither Lyell nor Adam Sedgwick, his instructor at Cambridge who took Darwin along on geological excursions to Wales, accepted Darwin's views on evolution. Though Sedgwick, a clergymen, appeared to resist evolution on theological grounds, Lyell's own initial opposition appears instead to have flowed from his long-standing opposition to catastrophism. He thought the fossil record showed there to be no net change, progressive or otherwise, in the basic composition of life.

Lyell delighted in the discovery of mammals in the Stonesfeld slates of England, which pushed their earliest occurrence way back to the Upper Triassic. He was convinced that ultimately mammals would be found to have lived in the Paleozoic. But it was a vain hope, as the earliest known mammals, after nearly a century and a half of further avid paleontological exploration, are still Upper Triassic in age.

Much the same is true of extinction: Huttonian-imbued cyclic steady-staters, opposed especially (and quite rightly) to the Bucklandian sort of catastrophism of the Noachian deluge, insisted that no species had truly become extinct. Thomas Jefferson, second President of the United States, was an amateur paleontologist who at one point filled the then-unfinished West Wing of the White House with Pleistocene mammalian fossils from Great Salt Lick, Kentucky. He was convinced that the sloths and mastodons, whose bones were among those assembled in his collection, would be ultimately found still alive and well in as-yet-uncharted realms of the continent.

18. Some Devonian organisms like conodonts occur in finer-scale zones; my 5- to 7-million-year estimate of Devonian zones actually refers to assemblages of many different kinds of invertebrate fossils discussed more fully in Chapter 6.

19. One reason that Murchison's work seemed to hold up better to scrutiny than Sedgwick's is that Murchison, with his dense fossil record, relied on fossils for his correlations. These were pre-evolutionary days: fossils were yet to be understood as remains of life that evolved and became extinct, leaving a linear record of change in the rocks. Therefore, it was not intrinsically clear that fossils were inherently any better than, say, mineral grains in making correlations among widespread outcrops. Thus, Sedgwick tried, unfortunately, to use mineralogy in his work.

Occasionally unique suites of minerals can serve as excellent markers for stratigraphic levels. (Volcanic ash beds are marvelous for precisely this use, not to mention the famous layer of iridium that left a worldwide fingerprint at the end of Cretaceous times.) But quartz, lime, and clay minerals—the most common constituents of sedimentary rocks—are timeless in the truest Huttonian sense: They form anew and are recycled, all in the same basic form and composition throughout the geological ages, and can tell us nothing about *when* they were formed.

Chapter 3

1. For recent discussions of the nature, extent, causes, and possible remedies of the current biodiversity crisis, see Wilson, 1993, and Eldredge, 1995b, 1998 and the many references cited therein. Indeed, it is partially because we are in the midst of a rapidly accelerating global extinction spasm that I feel it imperative to integrate the historically disparate strands of earth and biological science—the better to understand the nature of the global system and thus how to deal with this major predicament.

2. As a graduate student, I encountered the writings of one such observer whose name I have long since forgotten (a paleontologist as I recall). In a manuscript then circulating, this person wrote of the evident lack of connectivity among living forms, that is, species, or large-scale groupings of species. In effect this absence negated both sorts of patterns discussed in this section: one of inclusivity, the other of directionality. As to the Greeks, Aristotle, for one, classified organisms into various groups. He also propounded the linear scale of lower-to-higher organisms.

3. Lamarck bolstered his evolutionary argument-from-pattern with his own suggestions of the mechanisms of evolution: his use-and-disuse "inheritance of acquired characteristics," or the ability of organisms to pass along to descendants features they develop during their own lifetimes. But why did Darwin succeed in establishing the very idea of evolution where predecessors such as Lamarck had failed? Many say that Darwin succeeded because he supplied a mechanism (that is, natural selection) sufficiently plausible to explain how evolution actually occurs. But so, of course, did Lamarck. And there were few canons of inquiry or any empirical or experimental data to refute Lamarck's ideas on the inheritance of acquired characters. Indeed,

Darwin beat a retreat and resurrected distinctly Lamarckian notions in his sixth (and most widely read) edition of On the Origin of Species (1872). And it wasn't until August Weismann successfully propounded his distinction between germ plasm (sperm and eggs) versus the somatic line (nonreproductive cellular tissue) that biologists in general accepted the dictum that what happens to an organism during its lifetime cannot and will not be transmitted to its offspring—unless the "something" that happens causes mutations in the sex cells.

4. The simplest definition of Kingdom Animalia is that it consists of all heterotrophic multicellular organisms. Heterotrophs ingest other organisms for nutrients and energy; animals are distinct from certain microbial heterotrophs in that their bodies consist of many cells, and more than one kind of cell.

5. "Higher" and "lower" are generally considered pejoratives. As a graduate student in the 1960s, I found that it was even politically incorrect to use the word primitive in an evolutionary context, especially in conjunction with human evolution! But "simple" and "complex," and their relative degrees, are more value-free, more objectively ascertainable. Few will doubt that sponges are simpler than corals, or that corals are more simple than flatworms. The difference is in the degree of complexity. There is definitely a "signal" here: Within many of the major groups, and even within their subcomponent groups, there is a gradient from simpler to more complex forms.

6. Mayr (1982, p. 150 ff.) characterizes Aristotle's method of logic as dichotomous, that is, as described here for the invertebrate/vertebrate distinction. But he goes on to say that Aristotle's classification of animals (as opposed to his theoretical discussion on how to classify) was not itself dichotomous. Aristotle, in De partibus animalium, actually ridicules dichotomous classification.

7. Indeed, the biological world has universally embraced University of Massachusetts biologist Lynn Margulis's thesis, now some 25 years old, that eukaryotic cells are probably evolutionary hybrid fusions of several different sorts of bacterial prokaryotic cells.

8. The generic name is always given along with the specific, to avoid confusion. To my knowledge there is only one species with the name sapiens (that is, ourselves). But there are many with the name carolinensis (meaning "of the Carolinas"), used over and over again as the early naturalists encountered and described the fauna and flora of the New World. Sitta carolinensis, the white-breasted nuthatch, is a species of bird, while Sciurus carolinensis is the Carolina gray squirrel, a species which occurs over most of the eastern half of the North American continent.

9. Nelson was the central figure in bringing to the United States the methodology of phylogenetic systematics, codified and named by the German entomologist Willi Hennig (1966), and elaborated and promulgated by Hennig's "champion," the Swedish entomologist Lars Brundin, who was indeed more adept than Hennig himself in explaining what phylogenetic systematics is all about.

10. Darwin continues to sound fresh to my ears, prima facie evidence, perhaps, of his stature as the first of the truly modern scientists working in "natural history." Though I say that his goal was to convince his peers in the "thinking world" of the truth of his basic proposition that all organisms are descended from a single common ancestor, I was nonetheless struck by his candid remark, again in the very last

chapter (p. 423) that it is the younger generation(s), and not his contemporaries, whose minds were already made up on the issues, whom he most hoped to convince of the validity of his arguments. I have always felt the same way.

On another point, I use the word *fact* here with some reluctance. That the world is an oblate spheroid is also properly considered a fact, yet not too many centuries ago it was largely written off as a crazy idea, despite the otherwise convincing evidence for the earth being round established by the ancient Greeks. Facts are ideas, descriptions of the way things are, hypotheses that have withstood so many strong tests that their truth content is no longer in rational doubt. The proposition that all life is descended from a single common ancestor—the basic postulate of organic evolution—qualifies as a fact under this definition of the term.

11. Ernst Mayr (1991) has cribbed this apt self-description of the *Origin* as the title for his enlightening analysis of Darwin's thinking and his subsequent impact on the growth of evolutionary biology.

12. See especially Michael Ghiselin's excellent *The Triumph of the Darwinian Method* (1969) and Mayr's above-mentioned *One Long Argument* (1991) for detailed treatments of Darwin's mode of analytic thinking. These differ in important respects from my own emphasis on patterns in Darwin's presentation in the *Origin*.

13. Explication of the hypothetico-deductive method is most closely associated with physicist-turned-philosopher Karl Popper (e.g., 1959). It was Popper's difficulties with the notion of natural selection that apparently most influenced him to condemn evolutionary biology as outside the realm of scientific (meaning, hypothetico-deductive) method. But Popper, as a physicist, also held an experimentalist's eye on the nature of predictivity: that prediction implies the nature of events that have yet to happen. In contrast, all evolutionary biologists admit that they cannot predict the evolutionary future of species with any degree of confidence. Even if they could, no one would be around to see if the predictions came true. Missing in this line of thought is the simple realization that predictions may involve consequences that have already happened: The experiments have already been performed as life evolved. We predict patterns of evolutionary history that we ought to find if our suppositions about the nature of the evolutionary processes are correct. The philosopher Michael Scriven (1959) has referred to the prediction of past events as *retrodiction*.

14. Mayr (1942) explicitly devoted a passage to precisely the same problem Darwin posed with red grouse (that is, how to tell if two geographically disjunct "forms" are the same or a single species). Mayr remarked that confusion is inevitable given the very nature of the evolutionary process. Creating the confusion in the forms' classification is the expectation that, in a survey of the entire spectrum of species in the wild, there should be many examples of species caught in the act of splitting, of *speciating*.

15. Much has been made of Darwin's not noticing the differentiation patterns of the Galapagos finches while visiting the islands and collecting specimens as H.M.S. *Beagle*'s ship naturalist. Rather, it was not until ornithologist John Gould of the British Museum examined Darwin's collections that a pattern of differentiation into what are now recognized as some thirteen different species was picked up. Small wonder Darwin missed the radiation when he was there: The birds are invariably black and streaky, or a dull olive color. And the most striking adaptive change was a

remarkable piece of behavior: The "woodpecker" finch uses twigs to pry insects from the interstices of tree bark. Behavior doesn't accompany study skins back to the museum.

Since restudied intensively, most notably by David Lack (1947) and, most recently, by Peter and Rosemary Grant (see Grant, 1986), the Galapagos finches especially exemplify constant shifts in gene frequencies. As conditions, particularly rainfall patterns, shift from year to year, some plants are favored over others and consequently seeds of varying sizes and hardnesses are produced; the finches' beak configurations change to keep pace: Most of the "microevolution" in these finches involves fluctuations in bill shape and size. Grant (1986) reports statistical changes back and forth within a grand spectrum of variation in many of the species on each of the individual Galapagos islands. Natural selection seems to track these climatically induced fluctuations in seed production minutely by tinkering with beak proportions. Yet the change is not cumulative but rather static within a range of variation. For true differentiation to occur, it seems populations must spread to adjacent islands and diverge, just as Darwin originally proposed.

16. Many modern biogeographers (see, for example, Nelson and Rosen, 1981; Nelson and Platnick, 1981) have considered Darwin to have overstated his case for "centers of origins." This notion posited that ancestral species give rise to clusters of related species within a region, and the presence of one or more of those species outside that range necessarily implies a later episode of "dispersal." Thus, most dispersals would be expected to be isolated events, and the direction of dispersal more or less random. Yet patterns of relationship *between* floras and faunas of different regions are frequently observed to be more patterned—less random—than would be predicted under a purely dispersalist model such as Darwin's.

Thus, the regional fragmentation of biotas, which would be expected to lead to a differentiation of species in two regions formerly conjoined, is of course quite consistent with evolution—even the evolutionary patterns of divergence through isolation that Darwin had in mind. Nonetheless, he neither foresaw nor predicted such patterns. This is not surprising, given his overall reluctance to admit the possibility that numbers of new species, in different lineages, might appear at more or less the same time. Yet this is precisely what has happened since the Isthmus of Panama was completed some 3 million years ago. That event has led to the divergence of species formerly held in common in what are now the separate Caribbean and Pacific coasts of Panama (though biologists debate how much speciation has actually taken place). Such patterns are known as *vicariant;* closely related species living in adjacent regions are known as *vicars.* These vicars are seen as replacing one another and often performing similar ecological roles in one another's region. See the section "Pattern Denied" for more on Darwin's reluctance to consider patterns of coordinate evolution.

17. Comparative anatomists have been notorious, not only in their initial opposition to the very idea of evolution, but often their reluctance in later years to apply the concept directly and meaningfully to their research. Richard Owens, a central figure in mid-nineteenth-century British biology, was an outstanding case in point. His concept of *archetypes* (see Desmond, 1982) recognized a sort of least-common-denominator similarity in ground plan, linking up, for example, all the vertebrates in terms of their basic anatomical architecture. Archetypes did form a kind of

pattern, reflecting the progressively greater resemblances shared by vertebrate embryos the earlier in development they are compared. Nonetheless, they were essentially a *static* concept. For example, the greatly diversified vertebrates would thereby be construed as making up a welter of varying versions of that basic archetype. Archetypes were generally seen as created by God. Thus, they were a way of staving off the naturalistic, evolutionary interpretation of patterns of comparative anatomy. Modern creationists often adopt the very same view. Some of them see mammalian diversity, for example, as variant versions of a "created kind," perhaps even the result of "microevolutionary processes."

On the other hand, of course, many comparative anatomists quickly accepted the basic postulate of evolution. T. H. Huxley, in England, and the German August Haeckel took the practice of systematic arrangement of species into natural kinds and quickly converted it into a search for *phylogenetic pattern*—the genealogical relationships of species that Darwin had demonstrated nested Linnaean taxa to be really all about. The empirical pattern underlying the concept of "archetypes" was readily reinterpreted as "ancestors" or "common ancestors."

Still, patterns in nature can be studied in and of their own right. In the instance of nested sets of homologous characters or nested Linnaean taxa, many systematists adopting *cladistics* (*phylogenetic systematics;* see Chapter 1 and earlier discussion, this chapter) have taken a neutral stance. For example, they claim that they are indeed searching for natural kinds and not necessarily studying evolutionary history (see Nelson and Platnick, 1981). However, the vast majority of systematists, including cladists, follow the Darwin/Huxley/Haeckel lead: They see their search for nested patterns of homologous characters as the analysis of phylogenetic, or evolutionary, pathways.

18. The only competing paradigm, at least in the western world, that purports to explain biological diversity and interconnected similarity in nested sets of taxa is subsumed in the various forms of creationism. That a supernatural Creator may have fashioned life in this form, as posited by creationists, is a form of hypothesis beyond the purview of science. (For example, see Gish, 1978, who explicitly addresses nested sets of taxa from a creationist perspective.)

19. My sense of cross-genealogy here refers to the unrelated populations of species lineages located at a particular time and place (for example, the squirrels, oak trees, hawks, grasses, and fungi of my backyard woodlot). They interact to form the biotic component of what ecologists call local *ecosystems*.

20. This last sentence appearing in the first edition of the *Origin* is one of the few acknowledgments of *stasis* by Darwin. Darwin was to rectify this situation somewhat in the sixth edition of the *Origin*, in response to his critics.

21. Darwin, for good and sufficient reasons discussed here, was wholly occupied with genealogical pattern, and apparent exceptions, in order to establish the very idea of organic evolution. It may be that by his overlooking cross-genealogical patterns, the science of ecology got such a belated start—well into the twentieth century.

22. This is the very point of departure, and the prime significance of the achievements, of both the geneticist Theodosius Dobzhansky and the systematist Ernst Mayr, working in New York in the mid-1930s and early 1940s. (I will return to this subject in Chapter 5.) Suffice to say here that interspecific discontinuity to Darwin was a problem standing in the way of acceptance of his general thesis that new species arise from ancestral

species. To many of his successors, it became a pattern that evolutionary biology had to account for in a positive manner—and not simply explain away.

23. Ernst Mayr (Mayr, 1982) has been especially vocal and cogent on this point.

24. In his *Natural Selection: Domains, Levels, and Applications* (1992), evolutionary biologist George Williams is especially lucid on the distinction between corporeal entities such as genes, on the one hand, and the information they carry, on the other. I find much with which to disagree with Williams (spelled out in some detail in my *Reinventing Darwin*, 1995a), including the interpretation of this very distinction between genes and information. Nonetheless, the distinction itself is an extremely important one.

25. Genetic engineering is the latest manifestation of human manipulation of genetic information. Instead of breeding only those organisms selected from a variable population who best display the desired traits, genes identified as underlying those desired traits are introduced into organisms directly. For example, the "healthy" gene version, the *allele*, can be substituted for one that causes disease. However, unless such genes are introduced to the germ line (that is, sex cells), thus becoming potentially heritable, genetic engineering is no substitute or improvement on selective breeding, which, though working more slowly, directly changes the heritable (that is, gene-based) information of the stock.

26. I took the *Origin* to read after arriving early (to be sure of a seat) to a lecture given by Louis Leakey at Butler Library on the Columbia University Campus in 1964. I was worried that I wouldn't understand Victorian prose, and that the subject matter would prove too difficult. What a relief to find that Darwin exposed his ideas in a disarmingly clear and relatively simple way—this despite an undergraduate's initial unhappiness with Darwin's accounts as an acolyte pigeon fancier instead of cutting to the chase, as I then saw it.

27. That Darwin made this profound, and essentially correct, pronouncement without so much as a passing shudder is remarkable. Human population has climbed from between 1 and 10 million at the dawn of agriculturally based human existence ("civilization") to a full 6 billion at the millennium. See Joel Cohen's (1995) *How Many People Can the Earth Support?* for a cogent and sobering look at the history of human confrontation and calculations on our population growth. Biologists concerned with the current mass ("Sixth") extinction, most certainly including myself, see the sort of human population growth that Darwin mentions in passing as the main underlying destructive force.

28. The best-known example of contentious thinking in this field is the work of V. C. Wynne-Edwards, news of whose death has just reached me as I write this chapter (February 1997). Wynne-Edwards argued forcefully that in species with social organizations, the density and number of individuals are regulated by a host of factors pertaining to the very nature of the social interactions among component organisms. This led him to conclude that natural selection can act on entire populations as well as among individuals within populations. Darwin himself liberally peppered his own text with allusions to competition, hence selection, among populations and species. Thus he would not have been surprised, or perhaps even averse, to Wynne-Edwards's thinking. Nonetheless, conventional wisdom in evolutionary biological circles during Wynne-Edwards's career placed his emphasis on

the social-structure regulation of populations and group selection in the distinct minority.

29. See Hull, 1973.

30. See especially Dawkins, 1976, 1982.

31. I characterize such attitudes, and my difficulties with them, much more fully in my book *Reinventing Darwin* (1995a).

Chapter 4

1. In Chapter 3 we encountered biological patterns used extensively by Darwin simply to convince his reader that life has evolved. But the deep connections between such historical biological patterns and theories of evolutionary process is developed, in true Huttonian spirit, particularly in Chapters 5 and 6.

2. Suess, *The Face of the Earth*, 1905–1909. For a concise history of plate tectonics, from Suess through Taylor and Wegener, see Hallam, 1973, and Marvin, 1973. Here, I particularly rely on their interpretation of Suess's writings.

3. Kay published *North American Geosynclines* as Memoir 48 of the Geological Society of America in 1951. Dubbed by one graduate student "the secret message," for its supposed literary opacity, the book was a distillation of Kay's rich experiences studying the complex geologies of deformed belts. It examines his synthesis of vast amounts of information for his students in the classroom and as part of the war effort. (Kay was in charge of estimating the magnesium reserves in the United States during World War II.) The German geologist Hans Stille was engaged in a similar synthesis during World War II. Between the two of them, their picture of the origin and development of mobile belts, and the analogue with geological regions observable on the earth today, was the closest thing to a "theory of the earth" available right up into the 1960s. It was all the more remarkable that Kay, at a conference in Gander, Newfoundland, in 1968, arose at one point to ask how his older views on geosynclines could be modified to accommodate the no-longer-deniable fact that continents have been moving all over the landscape since the very beginnings of geological time.

4. My basic source on Wegener's arguments was Wegener, *The Origin of Continents and Oceans*, an English translation of the fourth revised edition of *Die Enstehung der Kontinente und Ozeane* (1929). According to Marvin (1973), the first edition of Wegener's book was "a slim volume of 94 pages," of which only two copies are to be found in U.S. libraries. It was not until the third edition (1922), translated from the German into several languages, including English, that Wegener's ideas became widely known—and debated. The fourth edition contains a maturation of Wegener's ideas on mechanism. They move away from tidal forces, toward notions of convection powered from within the earth. Such ideas are much more nearly similar to the later notions of plate tectonics. The fourth edition also includes replies to Wegener's (by then) many critics, much as did Darwin's sixth edition of *Origin*. Yet in that latter book, Darwin somewhat vitiated his argument, for example, by admitting the possibility of the inheritance of acquired characteristics. Whereas by all accounts Wegener sharpened and essentially strengthened his theory in his last edition.

Marvin, incidentally, traces Wegener's interest in the congruence of Atlantic coastlines back to 1903, when he was said to have pointed out the pattern to a classmate.

5. Wegener (1929; English translation 1966), p. 30.

6. Actually, Wegener (1929, p. 57) says Jeffreys's values of viscosity are "extreme"— that is, not in accordance with the views of his colleagues.

7. Wegener, 1929, pp. 69–70, quoting from DuToit (1927), *A Geological Comparison of South America with South Africa*, pp. 15–16. DuToit and Arthur Holmes were the two most influential supporters of Wegener's basic thesis of drifting continents. Geologists of the southern hemisphere, faced with the overwhelming evidence in favor of an ancient supercontinent, Gondwana, in general were always more receptive to the notion of continental drift than their colleagues to the north—especially, for reasons not entirely clear, those who lived in the United States.

8. Hallam, 1973.

9. To my bemusement, not to say chagrin, a symposium hosted by my own institution, The American Museum of Natural History, and featuring many of my intellectual heroes on the staffs of both the museum and my alma mater, Columbia University, stands out as perhaps the apotheosis of conservative, antidrift sentiment. There were a few brave dissenting voices, such as the invertebrate paleontologist Kenneth Caster (late of the University of Cincinnati and an expert, not surprisingly, on the paleontology of the southern hemisphere), but they were few and far between. Published as a *Bulletin of the American Museum of Natural History* in 1952, (with Ernst Mayr as editor) the volume makes fascinating reading. There is geophysicist Maurice Ewing, even then Director of Columbia's Lamont (now Lamont-Doherty) Geological Observatory, assuring everyone of the permanence of the continents and the oceans—on Jeffreysian geophysical grounds. Ewing was *still* director when Lamont geophysicists helped to lead the palace coup that virtually overnight turned the derided and discredited Wegenerian theory of continental drift into the glamorous, enthusiastically embraced theory of plate tectonics in the mid-1960s. Then, too, George Gaylord Simpson intones in sober terms the importance of land bridges and other means of animal and plant dispersal. He rejects all the fossil and recent biological evidence that Wegener earlier, and so many of us later, see as fundamentally sound evidence for the relative motions of the continents.

10. For example, Evans, Beukes, and Kirschvink, 1997, and Kirschvink, Ripperdan, and Evans, 1997.

Chapter 5

1. I'll never forget stumbling on an old, thick textbook on physiology late one evening at Columbia University's biology library in the mid-1960s. Though I have long since forgotten the book's title and even its author (I think the publication date was 1909), the message of the book's preface remains indelibly etched in my memory: With electricity finally in the lab, at last we can become true scientists and abandon the amateurish pursuits of antiquated, field-oriented natural history. The author was particularly unkind to Darwin, who seemed to symbolize old-fashioned natural history, with his penchant for discussing species and his reliance on ideas such as natural selection which were not experimentally confirmed.

As we shall see in this chapter, the discovery of genetics seemed to offer, to some turn-of-the-century biologists at least, an alternative to Darwinian natural selection. Indeed, the history of biology's progressive assault on the hidden physico-chemical microcosm of the cell—aided and abetted by ever-refined apparatus and experimental techniques—has been to repeatedly declare Darwinism dead. Supposedly, it has been replaced by the new understanding at the level of, first, the simple gene and, then, by its components once the nature and structure of DNA had become reasonably clear. Simple-minded reductionism in its most blatant form seems to go hand in glove with all major breakthroughs into the mysteries of the structure and chemistry of cells. Yet what ultimately happens is the recognition that whatever the new discoveries of the inner workings of cells and genes may be, they don't supplant the Darwinian vision. Instead, they must be integrated with the notion of natural selection. Why? For the very good reason that selection takes place at the level of populations, where it deals with patterns of variation and parental resemblance that arise from cellular mechanisms.

2. For example, experiments with flax (Cullis, 1988) and bacteria have suggested that mutations may arise as an ameliorating response to direct environmental stress, and they can be passed along in succeeding generations. And Thomas Cech has demonstrated the existence of *reverse transcriptase*, the key component of so-called retroviruses (such as HIV), which allows RNA to be assembled according to the structure of a protein is in direct contradiction to the central dogma. Cech's discovery has interesting implications for theoretical models on the origin of life. Such findings are the exceptions that literally "prove the rule." By providing the rare counterexamples, they actually highlight the near-ubiquity of process as Weismann and the central dogma conceive it. Both generalizations remain intact as the overwhelming pattern.

3. Mendel's experiments, leading to the "laws" of segregation and recombination, produced critical, telling ratios of various characteristics in his experimental pea populations. Some geneticists, going back as far as Bateson, have remarked on the overly pat neatness of Mendel's patterns, and some have even accused Mendel of fabricating his data. See Sapp, 1990, for an illuminating discussion.

4. Much of this work was carried out on the ninth floor of Schermerhorn Hall on the Columbia University campus in New York City, under the direction of biologist Thomas Hunt Morgan. Students and faculty associated with the famous "fly room" produced in short order paper after paper.

5. The German paleontologist W. Waagen (1868) had previously used the term *mutation* to denote subdivisions of lineages of Mesozoic ammonites. This term had not caught on in paleontology before its independent coinage in genetics decades later.

6. *Macromutations* are most closely associated with the name of Richard Goldschmidt, an outstanding geneticist of the mid-twentieth century. Unfortunately, he is largely remembered for his notion that such large-scale mutations might occasionally give rise to what he chose to call *hopeful monsters*. Yet the notion that relatively small mutational changes in the regulatory apparatus (the portion of the genetic machinery that regulates the expression of those genes that encode the structure of proteins) might have relatively large effects is an idea that continues to attract some evolutionary theorists.

7. Osborn's ego was as big as his fortune. My old vertebrate paleontology teacher, Edwin H. Colbert, was Osborn's assistant in the 1930s. Colbert tells the story of how Osborn, confronting a group of students issued into his august presence, put them at ease by telling them he knew exactly how they felt when, as a young man, he himself shook the hand of Thomas Henry Huxley. Thus he brought himself within one step of the great Charles Darwin himself.

8. See especially William Provine's 1986 study of Sewall Wright for details, not only of Wright's contributions, but his interactions with Fisher, who in many ways was his rival in the mathematization of evolutionary theory.

9. In a general discussion entitled "window dressing or genuine influence?", Mayr (1982, p. 851) makes the statement that the biologist William Ernest Castle, in 1903, "showed that the genotypic composition of a population remains constant when selection ceases, but [this result] was ignored until Hardy and Weinberg provided a mathematical formulation." This illustrates Mayr's contention that "a law, principle, or generalization" in biology is frequently overlooked or "ignored when first stated because it was phrased in words rather than in the form of a mathematical equation." Such a point is difficult to disagree with.

My own reaction to the Hardy-Weinberg equilibrium when first encountered was rather different. I was struck with the conflict between, on the one hand, the ambient feeling that evolution is inevitable given the simple passage of time, and, on the other, the certitude of genetic stability *unless* disrupted by immigration, mutation, selection, or non-random-mating behavior. That something has to happen to disrupt genetic stability was the abiding lesson I took away from Hardy-Weinberg.

10. In my *Reinventing Darwin*, I recount at length the current disputes between *ultra-Darwinists*, whose theory is imbued with population genetics, and evolutionary biologists, such as myself, known as *naturalists*. The latter group is eclectic, including paleontologists, systematists, ecologists, and others.

The ultra-Darwinian perspective maintains, at its very core, the Fisherian stance that natural selection is both necessary and sufficient to explain the evolution of life. Like any theory, it works best in the domain it was originally fashioned to explain, in this case, generation-by-generation adaptive change in gene frequencies. Other notable successes are in *sociobiology*, which explains patterns of cooperation in social organisms in terms of degree of shared genetic information among the participants. That system works especially well in the case of social insects. It was rumored to have been first worked out by Ronald Fisher on the back of an envelope in an English pub long ago. The sociobiological rubric works rather less well in other social organisms, such as birds and mammals (see Eldredge and Grene, 1992).

Occasionally, rather offhand comments have been made to the effect that the maximization of representation of one's own genetic material in the next generation will ultimately prove the key to understanding the structure of whole ecosystems. For example, this claim has been made by Richard Dawkins in *The Selfish Gene*. But on the whole this population genetics theoretical approach has failed to embrace cross-genealogical systems (for example, ecosystems), or make explicit connections with the dynamics of the physical, or abiotic, world. Partial exceptions to this sweeping condemnatory generalization of ultra-Darwinian evolutionary biology include the subject of co-evolution, in which two-species adaptive systems, such as butterfly/milkweed, are analyzed; Leigh Van Valen's notion of the Red Queen, where the evolution of one species is considered as a reaction to changes in other species impacting it; and innumerable nontheoretical population genetics studies in natural situations. It is especially the latter that continue to inform and lead to a more complete evolutionary theory.

For more on these disputes, see *Reinventing Darwin* and the references cited therein.

11. For example, the German biologist Moritz Wagner, the English biologist George Romanes, and the American David Starr Jordan, who later in his career was president of Stanford University.

12. Indeed, with the publication of Dobzhansky's *Genetics and the Origin of Species* in 1937, Columbia University Press turned an already outstanding series of books on biology into by far the most significant list on evolution. Included in the series were three editions of the latter work, Mayr's *Systematics and the Origin of Species* (1942), and Simpson's *Tempo and Mode in Evolution* (1944). All three were reprinted in recent years, Dobzhansky's and Mayr's under the "editorship" of myself and Stephen Jay Gould (we selected the titles and wrote new introductions), Simpson's with a new introduction by its author. Simpson, perhaps understandably, vastly preferred his own interpretation of his work to the essay I had written.

Additional important titles followed, and they still continue to appear in the Columbia list. For detailed analysis of the early works of Dobzhansky, Mayr, and Simpson, see my *Unfinished Synthesis* (Eldredge, 1985a), Chapters 2 and 3. For important insights on the genesis of Dobzhansky's *Genetics and the Origin of Species*, see William Provine's essay in *The Evolution of Theodosius Dobzhansky*, edited by Mark Adams (1994). For general treatments of the emergence of the modern synthesis, see especially Mayr and Provine's edited volume *The Evolutionary Synthesis: Perspectives on the Unification of Biology* (1980).

13. For example, the English biologist J. J. Romanes (1914) remarked that "without isolation, or the prevention of free intercrossing, evolution is in no case possible."

14. Mayr (1942) wryly points out that, if species be not real, then why have a theory of their origins? In a particularly eloquent passage (p. 152), he compares the process of speciation to the subdivision of one *Paramecium*, a single-celled organism. He declared that while a paramecium is in the process of dividing in two, it may be difficult to say if there is one or two paramecia; but once the division is complete, there are two individuals where once there had been but one. So, too, Mayr argued, with species: Species in the process of division confuse the picture. But once the division is complete, there are two discrete species where once there had been only one.

15. In sexually dimorphic species, the females of two species may resemble each other more closely than they do the males of their own species. Such is the case with two closely related hummingbird species of the North American west: Allen's hummingbird (*Selasphorus sasin*) and the rufus hummingbird (*Selasphorous rufus*). Taxonomists have always strived to put the right females with the right males in such situations, correcting past mistakes whenever breeding information has become available. This practice demonstrates that, despite definitions, interbreeding has always been considered an important attribute of species. Nonetheless, efforts to define what species are and how they evolve always focused far more on the anatomical features of individual organisms than on their reproductive behavior. That is, they did until Dobzhansky and Mayr turned things around by 180 degrees.

16. For example, Eldredge and Salthe, 1984; Eldredge, 1985a; Salthe, 1985.

17. Provine (1994, p. 112) quotes from an oral memoir left by Dobzhansky on the subject of his place in history and, in this brief excerpt, on his debt to Wright and his feelings about mathematics. They were not all that different from Mayr's: "Now . . . I certainly don't mean to make a preposterous claim that I have invented the synthetic or biological theory of evolution. It was, so to speak, in the air. . . . What that book of mine, however, did was—well, if you wish, in a sense, [popularize] this theory. Wright is very hard to read. He has a lot of abstruse, in fact almost esoteric, mathematics. Mathematics, incidentally, of a kind which I certainly do not claim to understand. I am not a mathematician at all. My way of reading Sewall Wright's papers, which I still think is perfectly defensible, is to examine the biological assumptions which the man is making, and to read the conclusion which he arrives at, and hope to goodness that what comes in between is correct."
   To all of this I can only say "Amen."

18. This is Wright's notion of *genetic drift*, a decidedly nondeterministic element to the regulation of gene frequencies sitting side by side with natural selection. Genetic drift was at first embraced by Dobzhansky and other evolutionary theorists, including, as we shall shortly see, George Gaylord Simpson. In the increasingly "hardened" version of the synthesis (see Gould, 1980, for use of this term), in the late 1940s and throughout the 1950s, selection was more often seen as the sole molder and shaper of the genetics of populations. And thus Wright became something of an outsider in the very evolutionary circles he had done so much to establish.

19. Wright, 1931.

20. Dobzhansky, 1951, pp. 9–10.

21. Mindful of all the subtleties of the problem though, in what I have called "Dobzhansky's dilemma," Dobzhansky also realized that, if selection focuses the genetically-based adaptive properties of a species too narrowly on an adaptive peak, there will be insufficient genetic variation on which selection can act should times change—should, in other words, the position of the "adaptive peak" change. One particularly important contribution of more recent analysts, especially George C. Williams, is to reassert the notion that selection cannot anticipate the future. To this day it remains an interesting and largely unaddressed conundrum that some species, the ecological generalists, are very variable, genetically and phenotypically, while others, the specialists, display very little variability. These alternative adaptive strategies generate typical, repeated patterns. Species whose organisms are ecological generalists tend to show very low rates of both speciation and extinction and tend to accrue very little evolutionary change over prodigious spans of time. Specialists, on the other hand, show high rates of speciation and extinction; their lineages tend to accumulate marked amounts of evolutionary change, often quite rapidly.

22. See, for example, Paterson, 1985.

23. I thank my mentor and colleague Bobb Schaeffer, now retired vertebrate paleontologist at the American Museum of Natural History, for showing me his photocopy of Simpson's response to this questionnaire.

24. Simpson himself alluded to the difficulties of producing *Tempo and Mode* during wartime, when he served in the U.S. Army. The book seems to have been written in at least two distinct phases as some peculiar, and I think rather telltale, incongruities in terminology seem to suggest. For example, he speaks of Wright's adaptive landscape, with its hills and valleys, in the beginning of the book. But he uses the term *adaptive zones* in the later chapters. Yet the intent seems the same, despite the linguistic shift. The book was started in the late 30s and literally interrupted by the advent of war, not appearing until 1944. For more details, see my essay on the structure and content of *Tempo and Mode* in Chapter 3 of *Unfinished Synthesis* (Eldredge, 1985a).

25. Such a measure, aptly named the *darwin*, was in fact subsequently devised by J. B. S. Haldane in 1949.

26. See Eldredge, 1984. The paper was published in a series of collected essays on the general subject of living fossils. The book was an attempt to examine common points, or *patterns*, linking disparate cases of famously slow, so-called arrested evolution.

27. Simpson actually used three terms in 1944: *micro-*, *macro-*, and *megaevolution*. He apologized for the latter term, which he introduced, and later dropped it. But in 1944, it was important for symmetry's sake to have three such terms, corresponding not only to his three distinct classes of evolutionary rates but also to his three "modes" of evolution: Microevolution goes with speciation. Macroevolution goes with evolution at the level of genera and perhaps families, and it pretty much fits into Simpson's phyletic mode of evolution. And megaevolution, involving shifts to new adaptive zones (as explained later in the book), corresponds to quantum evolution. Indeed, as further terminological evidence for the interrupted nature of the book's preparation, Simpson seems to substitute his three modes (speciation, phyletic, quantum evolution) for micro-, macro-, and megaevolution as used strictly in the earlier sections of the book.

28. Wright, 1945. This review was published, oddly enough, in the journal *Ecology*.

29. Papers presented at the meeting were published as *Genetics, Paleontology and Evolution* (Jepsen et al., editors, 1949). It marked an exciting, if short-lived, time when paleontologists and geneticists seemed really to have been talking to each other, and when paleontologists actively sought evolutionarily meaningful patterns (mostly pertaining to evolutionary rates) in their data. Paleontologists were also active participants in the earlier days of the Society for the Study of Evolution. But subsequent participation by paleontologists has waned drastically, though I am happy to note that I published the initial statement of what was soon to be dubbed *punctuated equilibria* in the society's journal *Evolution* (Eldredge, 1971).

30. See, for example, Russell Lande's paper in *Paleobiology* (Lande, 1986).

31. See my *Time Frames* (1985b; reprinted 1987) for a detailed history of the development of the idea of punctuated equilibria. See also *Reinventing Darwin* (1995a) for a more recent update on its current status.

32. Neither Dobzhansky, Mayr, nor Wright were paleontologists. Thus, they might be excused for ignoring stasis when evolutionary-minded paleontologists like Simpson were themselves content to ignore it. Yet I hold to my conviction that stasis is implied by Wright's very concept of the internal structure of species. It should

have been explicitly anticipated by evolutionary biologists starting with Wright himself in the early 1930s. For more on this, see my *Reinventing Darwin*, and further discussion below.

33. After initially dismissing punctuated equilibria, Mayr has claimed it for his own, based of course on our reliance of his work on species and speciation theory. Interestingly, both Wright and Simpson, for wholly different reasons, also claimed priority over Gould and myself, again after expressing initial opposition.

34. Specifically, from George Williams (1992).

35. The terms are not synonyms; see *Reinventing Darwin* for details. Paleontologist Elisabeth Vrba, whose insights led to the distinction between species selection and species sorting in the first place, has also suggested an alternative mechanism underlying the generation of evolutionary trends within evolutionary lineages involving many species. She calls this her *effect hypothesis*, which is also treated in some detail in my *Reinventing Darwin*.

Chapter 6

1. Indeed, evolutionary biologists tend to treat the fact that organisms are matter-energy transfer machines as "trivially true." Especially since the days of Watson and Crick in the early 1950s, all that really seems to matter to most biologists are the aspects of living systems that render them unique—namely, the genome. Even chemists often appear to feel this way. To many of them, the "origin of life," which obviously must entail an explanation of the origin of replicating nucleic acids, is too often seen as solely that issue. Not dealt with is the conjoined issue of the development of proteins and cellular structures along with nucleic-acid-based replication.

2. Recall that Simpson saw the job of fusion between the genetical and paleontological perspectives on evolution of his day as a difficult and even "dangerous" task. And I suppose it still is. The danger, of course, lies wholly in the realm of turf battles. Intellectually the task of fusing the perspective is difficult, but it is something all biologists ought to be trying to achieve. There is but one evolutionary history of life. And so, as Simpson said, and Darwin obviously saw, there must be a single theoretical structure that explains how life evolves.

3. Exceptions apparently include such important groups as spiders. According to my colleague Norman Platnick, spiders appear to reach their greatest diversity in the midlatitudes. Another exception that helps prove the rule is penguins and their northern-hemisphere ecological equivalents, the auks, murres, and puffins. They reach their highest levels of diversity in higher latitudes, becoming less diverse in lower latitudes.

4. In an excellent piece of scientific investigation, paleontologists Frank Stehli and colleagues (for example, Stehli et al., 1972), beginning in the mid-1960s, used Recent and geologically historical latitudinal diversity data to test the newly revivified hypothesis of continental drift/plate tectonics. Stehli's null hypothesis, that the continents had *not* changed position vis-à-vis one another, was accepted. Stehli found that, for example, Permian data (pertaining to the world around 450 million years ago) showed the same basic patterns of latitudinal diversity *when plotted on a map of the Recent world*. Thus he found himself an antidrifter as plate tectonics was

becoming the central paradigm of geology. Stehli found resolution of the conflicting results when it became clear that most post-Permian differential plate motion has been longitudinal rather than latitudinal.

5. As we have already seen, it is not entire species, but local populations of species, that occupy niches. Niches are defined as concerted roles played by local populations as parts of local ecosystems. Here, my casual equation of species diversity with niche diversity reflects merely semantic convenience. The actual mechanics relating species diversity to aspects of niche width and diversity work through the population, or *avatar*, level.

6. See my *Reinventing Darwin* (1995a, pp. 161–166) for a more complete discussion of this particular explanation for latitudinal diversity gradients. Contained therein are appropriate citations, especially the work of ecologist George Stevens.

7. The five are:
    (1) *Upper Ordovician*, generally considered to have come in two separate pulses of global cooling a million years apart (around 440 million years ago).
    (2) *Upper Devonian*, also commonly attributed to global cooling (around 367 million years ago).
    (3) *Permo-Triassic boundary* (245 million years ago), also generally interpreted as a protracted event of complex environmental causes, possibly including an anoxic upwelling and outpouring of carbon dioxide from the world's oceans. (However, as this book went to press, recent reports suggest that this extinction event may well have been caused by impact by an extraterrestrial object.) This, the greatest mass extinction that has yet occurred, may have taken out as many as 96 percent of the world's species, according to University of Chicago paleobiologist David M. Raup.
    (4) *Upper Triassic*, of debatable causes (208 million years ago).
    (5) *Cretaceous-Tertiary boundary* (65 million years ago). This mass extinction is arguably the most famous of all. It apparently consisted of many rapid-fire, closely spaced events, most likely a series of collisions between the earth and extraterrestrial objects such as comets or meteors. Or possibly it involved a reflection of massive volcanism on earth. This was the event that finally completely eliminated the terrestrial dinosaurs.
    As I have already remarked, many ecologists and systematists, myself included, are convinced that the modern world's species and ecosystems are currently in the throes of a human-induced Sixth Extinction. See my *Miner's Canary* (1991) for a survey of extinction phenomena in general, and my *Life in the Balance* (1998) for an overview of the present-day Sixth Extinction.

8. Norman Newell has been a leader in many areas of research, including the systematics and evolution of bivalved mollusks and the development of paleoecology where recent environments and animal communities are used as models for interpreting the past. He has also been an articulate spokesman for evolution in the perennial battles with creationists, and has served as mentor for many paleobiologists. (I am proud to have been one of his students, as was Stephen Jay Gould.) But perhaps Newell's most significant contribution has been his single-handed resurrection of cross-genealogical extinction and evolutionary rebound patterns. These he has painstakingly documented from the vast paleontological literature, and aptly termed them "crises" in the history of life (for example, Newell, 1967).

9. The only surviving remnant of the rhynchosaurs are two species of the lizardlike *Tuatara*, the "living fossils" of New Zealand.

10. I reiterate a point made earlier, in Chapter 1: Humans are utterly unique in no longer living inside local ecosystems. This revolution came as an immediate consequence of the development of agriculture some ten thousand years ago. Human population has gone from some 5 million then to over 6 billion now. The disruptive effects of so many people living on the planet, especially with the more recent advent of high technology, is the underlying cause of the mass extinction event currently plaguing the world's ecosystems and species.

11. Urbanization, highways, and especially agricultural transformation of the landscape have eliminated the ability of many species to continue to track habitats as climatic zones shift northward, in the present century, exacerbating their chances of becoming extinct.

12. An ecosystem consists of local populations of one or more (typically many more) species, plus the organic and inorganic nutrients, and all aspects of the local abiotic environment (substrate, climatic conditions, and so on), plus the all-important source of the energy that flows through the system. Solar radiation is the basis of energy in all ecosystems save deep-sea vent faunas and some simple archaebacterial systems. In both, energy is chemoautotrophically produced. In the former, energy is fixed by bacteria metabolizing sulfur in thermal vents, where the heat is the outcome of deep-seated radioactive decay. But in other ecosystems dependent on solar radiation, energy flows through the system primarily through patterns of production (for example, photosynthesis) and consumption (primarily herbivory, secondarily carnivory), plus the saprophytic decay feeding behavior of fungi.

13. A classic example was the overhunting of sea otters in the Pacific northwest. Their loss caused a population explosion of the sea urchins on which the otters preyed. The urchins then started overconsuming *their* food resource, the vast beds of kelp (a brown alga), which were the physical home to many invertebrate species. Removal of the sea otters caused the kelp ecosystem nearly to disappear in some places. With restoration of the otters, the kelp, with all their associated species, have recovered quite well.

14. Soft-bodied invertebrates, microbes, and algae are rarely preserved in such ancient fossil deposits. It is therefore not feasible to characterize collections of associated populations as ecosystems. Rather they are variously called *communities*, *assemblages*, or just old-fashioned *fauna*. Indeed, ecology itself is split into *community ecology* and *ecosystems ecology*, making all the more difficult my attempt here to associate Brett and Baird's *recurrent assemblages* (populations seen many times in different locations) with specific ecosystems. Consider the mosaic of Devonian seafloor life. It has a readily identifiable near-shore mollusk-dominated assemblage, an oxygen-poor algal and rhynchonellid brachiopod assemblage, and various other assemblages dominated by brachiopods and bryozoans, all of which occupy regular positions according to depth of water and, to a lesser extent, sediment type. These assemblages must reflect the structure of actual Devonian marine ecosystems, perhaps as identifiable subcomponents of the eastern North American Devonian inland seaway ecosystem.

15. Brett and Baird's work (see, for example, Brett and Baird, 1995) has already inspired a session at the national meeting of the Paleontological Society and its publication as an entire special issue, "New Perspectives on Faunal Stability in the Fossil Record" (1996), of the journal *Palaeogeography, Palaeoclimatology, Palaeoecology*. A valuable commentary on the entire coordinated stasis debate is given by paleontologist Arnold Miller (1997).

16. Not all paleobiologists and ecologists are fond of these types of patterns. I have already noted some of the disagreement surrounding coordinated stasis, though the disputes involve interpretations of the pattern rather than its basic nature. In contrast, some paleobiologists have even denied that the data from around 2.5 million years in eastern and southern Africa support the existence of the turnover pattern itself. But discussions with paleontologists familiar with eastern and southern African paleontology convince me the pattern is real. Biologist Paul Colinvaux (1996) has recently critiqued the refuge hypothesis as applied to tropical America. He gives it its due as a strong and immensely popular hypothesis that is, in good scientific practice, eminently testable. The hypothesis predicts, as Colinvaux states, that savannahs must have broken up the Amazonian forest during times of glacial maxima. But his review of the evidence convinces him that no such reticulation of forest and savannah occurred. Colinvaux does not deny the actual pattern of disjunct species distributions on which the hypothesis is based. However, he prefers to correlate such disjunctions with existing patterns of topographic relief and rainfall. The jury is still out.

17. See my *Reinventing Darwin* (1995a, Ch. 6) for a brief account of the development of the "ecological" and "genealogical" hierarchies—and their relationship.

18. James Lovelock's conception of *Gaia* is sometimes dismissed as an overly fanciful analogy that sees the whole earth, with its lithosphere, atmosphere, hydrosphere, and biosphere, as a megaversion of a single organism. The value of Gaia, however, lies in its specifications of the interconnectedness of these physical and biological components: the geochemical, biochemical, hydrological, and climatological cycles and interactions. These cycles and interactions do, in fact, render the surface of the planet, including ocean deeps and apparently also deep cracks in the earth's crust, as a single, dynamic system.

19. Darwin's concept of sexual selection is also clear from this diagrammatic depiction of the ecological and genealogical hierarchies. As we saw in Chapter 3, Darwin in 1871 defined sexual selection as the relative reproductive success within a single sex within a breeding population for reasons having solely to do with reproduction. This definition expressly excluded economic activity: Sexual selection is at the demal level of the genealogical hierarchy and has nothing to do with the economic hierarchy.

20. Here, in more succinct form, in which natural selection is explicitly linked to the patterns, is the "sloshing bucket" model of evolution. Each of the progressively higher-scale levels of the sloshing bucket retains all the lower levels. For example, consider natural selection: the effect of an organism's economic success on its reproductive success in the context of its life in both avatars and demes. It takes place at all levels of the sloshing bucket.

    Level 1. The organismal level is indeed the prime connection between the ecological (economic) and evolutionary (genealogical) domains. The economic lives of organisms affect their chances for reproduction (natural selection), while their reproductive activities replenish the players in the local ecosystem. Natural selection is generally stabilizing.

Level 2. Physically induced local ecosystem disturbance means the loss of avatars. Loss of avatars leads to recruitment from available demes—and the establishment of new avatars and the rebuilding of local ecosystems. This phenomenon is known as succession, the bailiwick of *Cecropia*. Natural selection takes place during succession, but it tends to be stabilizing, relying on evolutionary adaptations already in place for succession to proceed.

Level 3. Slow-acting physical change, (e.g. climatic change) leads to regional disruption of ecosystems, a larger-scale extension of ecological succession. Here again, natural selection goes on, as does recruitment from outlying avatars. Natural selection remains predominantly stabilizing, as habitat tracking is a pattern of movement to environmental conditions "recognized" by the pre-existing adaptations of individual species.

Level 4. Still higher on the scale are the "threshold" extinction effects, for example, turnover pulse. Here, true species extinction occurs, triggering true speciation. Natural selection, succession, and habitat tracking all occur. Selection is stabilizing for habitat-tracking species, strongly directional in speciation.

Level 5. In episodes of more nearly global devastation, so many species are driven to extinction that higher-scale taxa themselves become extinct. Succession and habitat tracking occur in the relatively few surviving species. And rampant speciation, with high survival rates for fledgling species and strongly directional natural selection, prompts radiations of lineages; hence the appearance, or at the very least, the blossoming, of higher taxa.

21. The salty water of the eastern Mediterranean sinks and flows westward under the incoming surface current of Atlantic waters. German submarines in World War II learned to cut their engines and ride these currents soundlessly in and out of the Mediterranean.

22. This information is from an article by Anthony G. Coates and Jorge Obando (1996) published in *Evolution and Environment in Tropical America*, edited by Jeremy B. C. Jackson, Ann F. Budd, and Anthony G. Coates. This important and stimulating collection of 13 papers (including that of Colinvaux, cited earlier in this chapter) offers detailed documentation of biotic and geological patterns, the correlations between them, and thoughtful analysis of the causal nexus of geotectonics, climate change, ecological disturbance, and evolution crucial to understanding the connections between the physical realm and biological evolution.

23. Jackson and Budd (1996), cited above, p. 5.

24. For an early analysis of the relationship between plate tectonics and evolution, see Cracraft, 1982.

Bibliography

Adams, M. B. (ed.) 1994. *The Evolution of Theodosius Dobzhansky*. Princeton University Press, Princeton.

Brady, R. H. 1994a. Explanation, description, and the meaning of "transformation" in taxonomic evidence. In R. W. Scotland, D. J. Siebert, and D. M. Williams (eds.), *Models in Phylogeny Reconstruction*. The Systematics Association Special Volume 52, pp. 11–29. Clarendon Press, Oxford.

Brady, R. H. 1994b. Pattern description, process explanation, and the history of morphological sciences. In *Interpreting the Hierarchy of Nature*, pp. 7–31. Academic Press, New York.

Brady, R. H. The idea in nature: Rereading Goethe's *Organics*. Unpublished manuscript.

Brett, C. E., and G. Baird. 1995. Coordinated stasis and evolutionary ecology of Silurian to Middle Devonian faunas in the Appalachian Basin. In R. Anstey and D. H. Erwin (eds.), *Speciation in the Fossil Record*, pp. 285–315. Columbia University Press, New York.

Bunge, M. 1977. *Treatise of Basic Philosophy*. Vol. 3: *The Furniture of the World*. D. Reidel Publishing Company, Dordrecht, Netherlands.

Clarke, J. M. 1889. The structure and development of the visual area in the trilobite *Phacops rana* Green. *Journal of Morphology* 2:253–271.

Coates, A. G., and J. A. Obando. 1996. The geologic evolution of the Central American Isthmus. In Jackson, J. B. C., A. F. Budd, and A. G. Coates (eds.), *Evolution and Environment in Tropical America*, pp. 1–20. University of Chicago Press, Chicago.

Cohen, J. E. 1995. *How Many People Can the Earth Support?* W. W. Norton, New York and London.

Colinvaux, P. A. 1996. Quarternary environmental history and forest diversity in the Neotropics. In J. B. C. Jackson, A. F. Budd, and A. G. Coates (eds.), *Evolution and Environment in Tropical America*, pp. 359–405. University of Chicago Press, Chicago.

Cracraft, J. 1982. A non-equilibrium theory for the rate-control of speciation and extinction and the origin of macroevolutionary patterns. *Systematic Zoology* 31:348–365.

Cullis, C. A. 1988. Control of variation in higher plants. In M.-W. Ho and S. W. Fox (eds.), *Evolutionary Processes and Metaphors*, pp. 49–61. John Wiley and Sons, New York.

Cuvier, G. 1812a. *Discours sur les Révolutions de la Surface du Globe*. Paris.

Cuvier, G. 1812b. *Recherches sur les Ossemens fossiles* 4 volumes, Paris.

Darwin, C. 1842. *The Structure and Distribution of Coral Reefs*. Reprint ed. 1984. University of Arizona Press, Tucson.

Darwin, C. 1859. *On the Origin of Species by Means of Natural Selection, or the Preservation of Favoured Races in the Struggle for Life*. John Murray, London.

Darwin, C. 1871. *The Descent of Man, and Selection in Relation to Sex*. John Murray, London.

Darwin, 1872. *On the Origin of Species*. 6th ed. John Murray, London.

Dawkins, R. 1976. *The Selfish Gene*. Oxford University Press, New York and Oxford.

Dawkins, R. 1982. *The Extended Phenotype: The Gene as the Unit of Selection*. W. H. Freeman and Co., Oxford and San Francisco.

Desmond, A. 1982. *Archetypes and Ancestors*. Blond & Briggs, London.

Dobzhansky, T. 1937. *Genetics and the Origin of Species*. Reprint ed., 1982. Columbia University Press, New York.

Dobzhansky, T. 1941. *Genetics and the Origin of Species*. 2d ed. Columbia University Press, New York.

Dobzhansky, T. 1951. *Genetics and the Origin of Species*. 3d ed. Columbia University Press, New York.

DuToit, A. L. 1927. *A Geological Comparison of South America with South Africa: With a Palaeontological Contribution by F. R. Cowper Reed*. Carnegie Inst. of Washington Publ. 381. Washington, D.C.

Eldredge, N. 1971. The allopatric model and phylogeny in Paleozoic invertebrates. *Evolution* 25:156–167.

Eldredge, N. 1984. Simpson's inverse: Bradytely and the phenomenon of living fossils. In N. Eldredge and S. M. Stanley (eds.), *Living Fossils*, pp. 272–277. Springer Verlag, New York.

Eldredge, N. 1985a. *Unfinished Synthesis: Biological Hierarchies and Modern Evolutionary Thought*. Oxford University Press, New York.

Eldredge, N. 1985b. *Time Frames*. Simon and Schuster, New York.

Eldredge, N. 1986. Information, economics and evolution. *Ann. Rev. Ecol. Syst.* 17:351–369.

Eldredge, N. 1989. *Macroevolutionary Patterns and Evolutionary Dynamics: Species, Niches, and Adaptive Peaks*. McGraw-Hill, New York.

Eldredge, N. 1991. *The Miner's Canary: Unraveling the Mysteries of Extinction*. Prentice-Hall Books, New York.

Eldredge, N. 1993. History, function and evolutionary biology. *Evolutionary Biology* 27:33–50.

Eldredge, N. 1995a. *Reinventing Darwin: The Great Debate at the High Table of Evolutionary Theory*. John Wiley and Sons, New York.

Eldredge, N. 1995b. *Dominion*. Henry Holt and Co., New York.

Eldredge, N. 1998. *Life in the Balance: Humanity and the Biodiversity Crisis*. Princeton University Press, Princeton, NJ.

Eldredge, N., and S. J. Gould. 1972. Punctuated equilibria: An alternative to phyletic gradualism. In T. J. M. Schopf (ed.), *Models in Paleobiology*, pp. 82–115. Freeman, Cooper, San Francisco.

Eldredge, N., and M. Grene. 1992. *Interactions: The Biological Context of Social Systems*. Columbia University Press, New York.

Eldredge, N., and S. N. Salthe. 1984. Hierarchy and evolution. *Oxford Survs. Evol. Biol.* 1:182–206.

Evans, D. A., N. J. Beukes, and J. L. Kirschvink. 1997. Low-latitude glaciation in the Palaeoproterozoic era. *Nature* 386:262–266.

Ghiselin, M. T. 1969. *The Triumph of the Darwinian Method*. University of California Press, Berkeley and Los Angeles.

Ghiselin, M. T. 1987. Species concepts, individuality, and objectivity. *Biol. and Phil.* 2:127–143.

Gish, D. T. 1978. *Evolution: The Fossils Say No!* Creation Life Publishers.

Gould, S. J. 1965. Is uniformitarianism necessary? *Am. J. Sci.* 263:223–228.

Gould, S. J. 1980. G. G. Simpson, paleontology, and the modern synthesis. In E. Mayr and W. B. Provine (eds.), *The Evolutionary Synthesis: Perspectives on the Unification of Biology*, pp. 153–172. Harvard University Press, Cambridge, MA.

Grant, P. R. 1986. *Ecology and Evolution of Darwin's Finches*. Princeton University Press, Princeton, NJ.

Grene, M. J. 1987. Hierarchies in biology. *American Scientist* 75:504–510.

Hall, J. 1859. *Palaeontology of New York*. Vol. 3. Albany, NY.

Hallam, A. 1973. *A Revolution in the Earth Sciences: From Continental Drift to Plate Tectonics*. Oxford University Press, Oxford.

Hamilton, W. D. 1964a. The genetical evolution of social behavior, I. *J. Theor. Biol.* 7:1–16.

Hamilton, W. D. 1964b. The genetical evolution of social behavior, II. *J. Theor. Biol.* 7:17–52.

Hamilton, W. D. 1996. *Narrow Roads of Gene Land*. Vol. 1: *Evolution of Social Behaviour*. W. H. Freeman, New York.

Hanson, N. R. 1958. *Patterns of Discovery*. Cambridge University Press, Cambridge, England.

Hennig, W. 1966. *Phylogenetic Systematics*. University of Illinois Press, Urbana.

Holmes, A. 1965. *Principles of Physical Geology*. 2d ed. The Ronald Press Company, New York.

Horgan, J. 1996. *The End of Science*. Addison Wesley, Reading, MA.

Hull, D. L. 1973. *Darwin and His Critics*. Harvard University Press, Cambridge, MA.

Hutton, J. 1788. *Theory of the Earth*. *Transactions Royal Society of Edinburgh*, 1:209–304.

Hutton, J. 1795. *Theory of the Earth, with Proofs and Illustrations*. Edinburgh.

Jackson, J. B. C., and A. F. Budd. 1996. Evolution and environment: Introduction and overview. In J. B. C. Jackson, A. F. Budd, and A. G. Coates (eds.), *Evolution and Environment in Tropical America*, pp. 1–20. University of Chicago Press, Chicago.

Jackson, J. B. C., A. F. Budd, and A. G. Coates (eds.). 1996. *Evolution and Environment in Tropical America*. University of Chicago Press, Chicago.

Jepsen, G. L., G. G. Simpson, and E. Mayr (eds.). 1949. *Genetics, Paleontology and Evolution*. Princeton University Press, Princeton, NJ.

Lyell, C. 1830–1833. *Principles of Geology*, 3 vols. London.

Kay, M. 1951. *North American Geosynclines: Geological Society of America Memoir* 48. New York.

Kirschvink, J. L., R. L. Ripperdan, and D. A. Evans. 1997. Evidence for a large-scale reorganization of Early Cambrian continental masses by inertial interchange true polar wander. *Science* 277:541–545.

Kricher, J. C. 1989. *A Neotropical Companion*. Princeton University Press, Princeton, NJ.

Kuhn, T. S. 1962. *The Structure of Scientific Revolutions*. University of Chicago Press, Chicago.

Lack, D. 1947. *Darwin's Finches*. Cambridge University Press, Cambridge, England.

Lande, R. 1986. The dynamics of peak shifts and the pattern of morphological evolution. *Paleobiology* 12:343–354.

Lankester, E. R. 1881. *Limulus*, an arachnid. *Quarterly Jour. Micro. Sci.* 21:504–548.

Lieberman, B. S., C. E. Brett, and N. Eldredge. 1995. A study of stasis and change in two species lineages from the Middle Devonian of New York state. *Paleobiology* 21:15–27.

Linnaeus, C. 1758. *Systema naturae*. 10th ed. Stockholm.

Malthus, T. R. 1826. *An Essay on the Principle of Population*. Murray, London.

Marvin, U. 1973. *Continental Drift: The Evolution of a Concept*. Smithsonian Institution Press, Washington, D.C.

Mayr, E. 1942. *Systematics and the Origin of Species*. Reprint ed., 1982. Columbia University Press, New York.

Mayr, E. (ed.) 1952. The problem of land connections across the South Atlantic, with special reference to the Mesozoic. *Bull. American Museum of Natural History* 99(3):79–258.

Mayr, E. 1961. Cause and effect in biology. *Science* 134:1501–1506.

Mayr, E. 1982. *The Growth of Biological Thought*. Harvard University Press, Cambridge, MA.

Mayr, E. 1991. *One Long Argument: Charles Darwin and the Genesis of Modern Evolutionary Thought*. Harvard University Press, Cambridge, MA.

Mayr, E., and W. B. Provine (eds). 1980. *The Evolutionary Synthesis: Perspectives on the Unification of Biology*. Harvard University Press, Cambridge, MA.

Miller, A. I. 1997. Coordinated stasis or coincident relative stability? *Paleobiology* 23:155–164.

Nagel, E. 1961. *The Structure of Science*. Harcourt, Brace, New York.

Nelson, G. J., and N. I. Platnick. 1981. *Systematics and Biogeography: Cladistics and Vicariance*. Columbia University Press, New York.

Newell, N. D. 1967. Revolutions in the history of life. *Geol. Soc. Amer. Spec. Paper* 89:63–91.

Paterson, H. E. H. 1985. The recognition concept of species. In E. S. Vrba (ed.), *Species and Speciation*. Transvaal Mus. Monogr. 4:21–29.

Platnick, N. I., and H. D. Cameron. 1977. Cladistic methods in textual, linguistic, and phylogenetic analysis. *Syst. Zool.* 26:380–385.

Playfair, J. 1802. *Illustrations of the Huttonian Theory of the Earth*. Edinburgh. 1956 reprint, Dover, New York.

Popper, K. R. 1959. *The Logic of Scientific Discovery*. Harper Torchbooks, New York.

Provine, W. B. 1986. *Sewall Wright and Evolutionary Biology*. University of Chicago Press, Chicago.

Provine, W. B. 1994. The origin of Dobzhansky's *Genetics and the Origin of Species*. In M. B. Adams (ed.), *The Evolution of Theodosius Dobzhansky*, pp. 99–114. Princeton University Press, Princeton, NJ.

Romanes, J. J. 1914. *Darwin and After Darwin. Vol. 3: Post-Darwinian Questions, Isolation and Physiological Selection*. The Open Court Publishing Company, Chicago and London.

Rudwick, M. J. S. 1985. *The Great Devonian Controversy*. University of Chicago Press, Chicago.

Salthe, S. N. 1985. *Evolving Hierarchical Systems*. Columbia University Press, New York.

Sapp, J. 1990. The nine lives of Gregor Mendel. In H. E. LeGrand (ed.), *Experimental Inquiries*, pp. 137–166. Kluwer Academic Publishers, Amsterdam.

Scriven, M. 1959. Explanation and prediction in evolutionary theory. *Science* 130:477–482.

Simon, H. A. 1962. The architecture of complexity. *Proc. Amer. Phil. Soc.* 106:467–482.

Simpson, G. G. 1944. *Tempo and Mode in Evolution*. Columbia University Press, New York.

Simpson, G. G. 1953. *The Major Features of Evolution*. Columbia University Press, New York.

Simpson, G. G. 1961. *Principles of Animal Taxonomy*. Columbia University Press, New York.

Simpson, G. G. 1963. Historical science. In C. C. Albritton, Jr. (ed.), *The Fabric of Geology*, pp. 24–48. Freeman, Cooper and Co., Stanford.

Simpson, G. G., and Roe, A. 1939. *Quantitative Zoology*. McGraw-Hill, New York.

Simpson, G. G., A. Roe, and Lewontin, R. C. 1960. *Quantitative Zoology*. Rev. Ed. Harcourt, Brace, New York.

Stehli, F. G., R. G. Douglas, and I. A. Kafescioglu. 1972. Models for the evolution of planktonic foraminifera. In T. J. M. Schopf (ed.), *Models in Paleobiology*, pp. 116–128. Freeman, Cooper, San Francisco.

Suess, E. 1905–1909. *The Face of the Earth*, Oxford University Press, Oxford.

Vrba, E. S. 1985. Environment and evolution: Alternative causes of the temporal distribution of evolutionary events. *S. Afr. J. Sci.* 81:229–236.

Wegener, A. 1929 (1966). *The Origin of Continents and Oceans*. Trans. of *Die Enstehung der Kontinente und Ozeane*, 4th ed. (1929). Dover, New York.

Whewell, W. 1837. *History of the Inductive Sciences*. Parker, London.

Williams, G. C. 1966. *Adaptation and Natural Selection: A Critique of Some Current Evolutionary Thought*. Princeton University Press, Princeton, NJ.

Williams, G. C. 1992. *Natural Selection: Domains, Levels, and Applications*. Oxford University Press, New York.

Wilson, E. O. 1993. *The Diversity of Life*. Harvard University Press, Cambridge, MA.

Wimsatt, W. C. 1980. Reductionist research strategies and their biases in the units of selection controversy. In T. Nickles (ed.), *Scientific Discovery: Case Studies*, pp. 213–259. D. Riedle Publishing Company, Dordrecht, Netherlands.

Wright, S. 1931. Evolution in Mendelian populations. *Genetics* 16:97–159.

Wright, S. 1932. The roles of mutation, inbreeding, crossbreeding, and selection in evolution. *Proc. Sixth Int. Congr. Genetics* 1:356–366.

Wright, S. 1945. *Tempo and Mode in Evolution*: A critical review. *Ecology* 26: 415–419.

Index